Material Politics

RGS-IBG Book Series

Published

Material Politics: Disputes Along the Pipeline
Andrew Barry

Everyday Moral Economies: Food, Politics and Scale in Cuba
Marisa Wilson

Working Lives – Gender, Migration and Employment in Britain, 1945–2007
Linda McDowell

Fashioning Globalisation: New Zealand Design, Working Women and the Cultural Economy
Maureen Molloy and Wendy Larner

Dunes: Dynamics, Morphology and Geological History
Andrew Warren

Spatial Politics: Essays for Doreen Massey
Edited by David Featherstone and Joe Painter

The Improvised State: Sovereignty, Performance and Agency in Dayton Bosnia
Alex Jeffrey

Learning the City: Knowledge and Translocal Assemblage
Colin McFarlane

Globalizing Responsibility: The Political Rationalities of Ethical Consumption
Clive Barnett, Paul Cloke, Nick Clarke and Alice Malpass

Domesticating Neo-Liberalism: Spaces of Economic Practice and Social Reproduction in Post-Socialist Cities
Alison Stenning, Adrian Smith, Alena Rochovská and Dariusz Świątek

Swept Up Lives? Re-envisioning the Homeless City
Paul Cloke, Jon May and Sarah Johnsen

Aerial Life: Spaces, Mobilities, Affects
Peter Adey

Millionaire Migrants: Trans-Pacific Life Lines
David Ley

State, Science and the Skies: Governmentalities of the British Atmosphere
Mark Whitehead

Complex Locations: Women's Geographical Work in the UK 1850–1970
Avril Maddrell

Value Chain Struggles: Institutions and Governance in the Plantation Districts of South India
Jeff Neilson and Bill Pritchard

Queer Visibilities: Space, Identity and Interaction in Cape Town
Andrew Tucker

Arsenic Pollution: A Global Synthesis
Peter Ravenscroft, Hugh Brammer and Keith Richards

Resistance, Space and Political Identities: The Making of Counter-Global Networks
David Featherstone

Mental Health and Social Space: Towards Inclusionary Geographies?
Hester Parr

Climate and Society in Colonial Mexico: A Study in Vulnerability
Georgina H. Endfield

Geochemical Sediments and Landscapes
Edited by David J. Nash and Sue J. McLaren

Driving Spaces: A Cultural-Historical Geography of England's M1 Motorway
Peter Merriman

Badlands of the Republic: Space, Politics and Urban Policy
Mustafa Dikeç

Geomorphology of Upland Peat: Erosion, Form and Landscape Change
Martin Evans and Jeff Warburton

Spaces of Colonialism: Delhi's Urban Governmentalities
Stephen Legg

People/States/Territories
Rhys Jones

Publics and the City
Kurt Iveson

After the Three Italies: Wealth, Inequality and Industrial Change
Mick Dunford and Lidia Greco

Putting Workfare in Place
Peter Sunley, Ron Martin and Corinne Nativel

Domicile and Diaspora
Alison Blunt

Geographies and Moralities
Edited by Roger Lee and David M. Smith

Military Geographies
Rachel Woodward

A New Deal for Transport?
Edited by Iain Docherty and Jon Shaw

Geographies of British Modernity
Edited by David Gilbert, David Matless and Brian Short

Lost Geographies of Power
John Allen

Globalizing South China
Carolyn L. Cartier

Geomorphological Processes and Landscape Change: Britain in the Last 1000 Years
Edited by David L. Higgitt and E. Mark Lee

Forthcoming

Smoking Geographies: Space, Place and Tobacco
Ross Barnett, Graham Moon, Jamie Pearce, Lee Thompson and Liz Twigg

Peopling Immigration Control: Geographies of Governing and Activism in the British Asylum System
Nick Gill

Geopolitics and Expertise: Knowledge and Authority in European Diplomacy
Merje Kuus

The Geopolitics of Expertise In the Nature of Landscape: Cultural Geography on the Norfolk Broads
David Matless

Rehearsing the State: The Political Practices of the Tibetan Government-in-Exile
Fiona McConnell

Frontier Regions of Marketization: Agribusiness, Farmers, and the Precarious Making of Global Connections in West Africa
Stefan Ouma

Articulations of Capital: Global Production Networks and Regional Transformations
John Pickles, Adrian Smith & Robert Begg, with Milan Buček, Rudolf Pástor and Poli Roukova

Origination: The Geographies of Brands and Branding
Andy Pike

Making Other Worlds: Agency and Interaction in Environmental Change
John Wainwright

Material Politics

Disputes Along the Pipeline

Andrew Barry

WILEY Blackwell

This edition first published 2013
© 2013 John Wiley & Sons, Ltd

Registered Office
John Wiley & Sons, Ltd, The Atrium, Southern Gate, Chichester, West Sussex,
PO19 8SQ, UK

Editorial Offices
350 Main Street, Malden, MA 02148-5020, USA
9600 Garsington Road, Oxford, OX4 2DQ, UK
The Atrium, Southern Gate, Chichester, West Sussex, PO19 8SQ, UK

For details of our global editorial offices, for customer services, and for information about
how to apply for permission to reuse the copyright material in this book please see our
website at www.wiley.com/wiley-blackwell.

The right of Andrew Barry to be identified as the author of this work has been asserted
in accordance with the UK Copyright, Designs and Patents Act 1988.

Library of Congress Cataloging-in-Publication Data
Barry, Andrew, 1960–
    Material politics : disputes along the pipeline / Andrew Barry.
      pages   cm
    Includes index.
    ISBN 978-1-118-52911-9 (hardback) – ISBN 978-1-118-52912-6 (paper)
1. Geopolitics.   2. Materials–Political aspects.   3. Material culture–Political
aspects.   4. Petroleum pipelines–Political aspects.   5. International economic
relations–Political aspects.   I. Title.
    JC319.B39 2013
    320.1′2–dc23
                                                              2013018241
A catalogue record for this book is available from the British Library.

Cover image: © Andrew Barry
Cover design by Workhaus

Set in 10/12pt Plantin by SPi Publisher Services, Pondicherry, India

1   2013

For Georgie

Contents

Series Editors' Preface

The RGS-IBG Book Series only publishes work of the highest international standing. Its emphasis is on distinctive new developments in human and physical geography, although it is also open to contributions from cognate disciplines whose interests overlap with those of geographers. The Series places strong emphasis on theoretically-informed and empirically-strong texts. Reflecting the vibrant and diverse theoretical and empirical agendas that characterize the contemporary discipline, contributions are expected to inform, challenge and stimulate the reader. Overall, the RGS-IBG Book Series seeks to promote scholarly publications that leave an intellectual mark and change the way readers think about particular issues, methods or theories.

For details on how to submit a proposal please visit:
www.rgsbookseries.com

Neil Coe
National University of Singapore

Joanna Bullard
Loughborough University, UK

RGS-IBG Book Series Editors

Figures and Tables

Acknowledgements

This book reports on a huge body of social research carried out along the route of a 1760 km pipeline. Much of the book is based on my own reading of the reports that this enterprise generated, whether they derived from the work of consultants, officials or NGOs. When weighed up against this research, my own fieldwork along the route of the pipeline through Georgia was modest in scale, but it also proved highly productive. For this I am grateful to my colleagues, assistants and informants. I am particularly indebted to my researcher Joanna Ewart-James, whose organisational ability and good sense made an enormous contribution to the success of six months of fieldwork in the UK and Georgia in early 2004. While in Tbilisi, we were initially assisted by the staff of the British Council. My thanks go, in particular, to Jo Bakowski, then Director of the British Council in Georgia, and to Louis Plowden-Wardlaw for introducing us. Tamta Khalvashi and Alex Scrivener provided invaluable assistance and translation skills as well as their knowledge of, and insights into, Georgian political history. Alex's good humour and tolerance of my determination to track down seemingly obscure details made fieldwork with him a real pleasure. I also owe a debt of gratitude to Farideh Heyat, who carried out a short period of fieldwork on the development of BTC in Azerbaijan on my behalf, which was the starting point for my account of transparency. While much of this book is based is based on evidence found in public documents, the BTC pipeline was also the subject of two important documentary films. I am especially grateful to Marin Skalsky for a series of conversations about his film *Zdroj* and to Nino Kirtadze for sending me a copy of her *Un Dragon dans les eaux pures du Caucase* at a timely moment.

This project had multiple origins. Meltem Ahiska's research on the history of Turkish radio broadcasting prompted me to think about the importance of archives and the formation of publics. James Marriott from

Platform provided invaluable suggestions and contacts during the early stages of the project. Barry Halton and Elizabeth Wild from BP gave me their support, and helped me arrange initial meetings with BP staff working in offices in Tbilisi, Baku, Ankara and London. I am very grateful to the employees of BP in Tbilisi, who were open about some of the challenges that they faced, always professional in their conduct, and generous with their time. During the course of my research in Georgia I benefited enormously from a series of interviews with Manana Kochladze from Green Alternative. Through the help of both Manana and BP staff in Tbilisi I was able to accompany Georgian NGOs and BP Community Liaison Officers to a number of the villages along the pipeline. My thanks also to the many informants I interviewed in London, Washington, Ankara, Tbilisi and Baku who are not named in the text as well as others who assisted me. None of my informants sought to impose conditions on the conduct of my research or its publication. Fieldwork in the region was funded by a grant from the UK Economic and Social Research Council *Science in Society* programme ('Social and Human Rights Impact Assessment and the Governance of Technology', 2003–4, RES-151-25-0011).

The idea for this book was first conceived in 2005, when I was working at Goldsmiths College in London. My thanks go to Celia Lury and Mariam Motamedi-Fraser, who co-founded the Centre for the Study of Invention and Social Process at Goldsmiths with me, for their friendship and inspiration. My thanks also to Tim Mitchell and the members of International Center for Advanced Studies and NYU in 2004–5 for their interest and support, and to Soumhya Venkatesen for lending me her room. More recently, I was lucky enough to work closely with Derek McCormack, Richard Powell and Ali Rogers, whose example helped me to understand what it means to think geographically. My thanks also to Ailsa Allen for the care that she took in preparing illustrations for the book, to my students at St Catherine's College for forcing me to think about the relevance of my concerns to theirs, and to Theo Barry-Born for his editorial assistance. Earlier drafts of many of the chapters were read by a number of colleagues and friends. I am enormously grateful to Lionel Bently, Bruce Braun, Mick Halewood, Caroline Humphrey, Jamie Lorimer, Andy Stirling, Gisa Weszkalnys and Sarah Whatmore and, above all, to Tom Osborne and Corin Throsby. In the latter stages of the development of the manuscript, I was also fortunate to receive insightful comments from the editorial board of the RGS-IBG book series, as well as an astute report from an anonymous reviewer. I remain grateful to Steve Hinchliffe for his encouragement, Neil Coe for his timely and clear editorial recommendations, and to Jacqueline Scott for being such an effective editor. Anne Piper provided an ideal refuge in Wytham over several years.

The book would not have been possible without the support of Theo and Clara and, above all, Georgie Born. Georgie's own research powerfully informed my sense of the value and the practice of ethnography. She has been a constant interlocutor throughout the development of this book and a critical and careful reader. It is therefore dedicated to her, with my gratitude and love.

Oxford, March 2013

Earlier versions of the following chapters have been published previously. Permission to publish this material is gratefully acknowledged.

Chapter 3 – Akrich, M., Barthe, Y., Muniesa, F. and Mustar, F. (eds) (2010) *Débordements: Mélanges Offert à Michel Callon*, Paris: Presses des Mines, 21–40.

Chapter 7 – Braun, B. and Whatmore, S. (eds) (2010) *Political Matter: Technoscience, Democracy, and Public Life*, Minneapolis, MN: Minnesota University Press, 89–118.

Abbreviations

ACG	Azeri-Chirag-Guneshli oil field
AGI	Above ground installation
APG	Artist Placement Group
APLR	Georgian Association for Protection of Landowners Rights
BPEO	Best Practicable Environmental Option
BTC	Baku-Tbilisi-Ceyhan Pipeline
CAO	Compliance Advisor Ombudsman
CDAP	Caspian Development Advisory Panel
CDI	Community Development Initiative
CEE	Central and East European Bankwatch Network
CIP	Community Investment Programme
DFID	UK government Department for International Development
DEAO	District Executive Authorities Office
DSA	Designated State Authority
ECGD	UK government Export Credits Guarantee Department
EBRD	European Bank for Reconstruction and Development
EIA	Environmental Impact Assessment
EITI	Extractive Industries Transparency Initiative
ERM	Environmental Resources Management
ESAP	Environmental and Social Action Plan
ESIA	Environmental and Social Impact Assessment
ESM	Environment and Social Management Plan
ESR	Environmental and Social Report to Lenders
FFM	Fact-Finding Mission
GIOC	Georgian International Oil Corporation
GSSOP	Georgia Sustainment and Stability Operations Program
GTEP	Georgia Train and Equip Program
GYLA	Georgian Young Lawyers Association
HDD	Horizontal Directional Drilling

HGA	Host Government Agreement
IAT	Indicator Assessment Tool
IDP	Internally displaced person
IEC	Independent Environmental Consultants Report to Lenders
IFC	International Finance Corporation
IFI	International Financial Institutions
IRM	Independent Recourse Mechanism
KHRP	Kurdish Human Rights Project
MSG	Multi-stakeholder group
NCEIA	Netherlands Commission for Environmental Impact Assessment
OECD	Organisation for Economic Co-operation and Development
OPEC	Organization of the Petroleum Exporting Countries
OSCE	Organisation for Security and Co-operation in Europe
OSR	Oil Spill Response Plan
PCIP	Project Community Investment Plan
PCDP	Public Consultation and Disclosure Plan
PKK	Kurdistan Workers' Party
PMDI	Pipeline Monitoring and Dialogue Initiative
RAP	Resettlement Action Plan
RR	Regional Review
SCP	South Caucasus Pipeline
SLRF	state land replacement fee
SOCAR	State Oil Company of Azerbaijan Republic
SPJV	Spie Capag/Petrofac International Joint Venture
SR	Sustainability Report
SRAP	Social and Resettlement Action Plan Review
STP	São Tomé e Príncipe
STS	Science and technology studies
UNDP	United Nations Development Programme
UNECE	United Nations Economic Commission for Europe
UKNCP	United Kingdom National Contact Point for the OECD guidelines
WWF	World Wildlife Fund

Chapter One
Introduction

In July 2004 officials from the International Finance Corporation (IFC) visited the small village of Dgvari, in the mountains of the Lesser Caucasus, in the region of the spa town of Borjomi in Western Georgia. The village, which was built on a slope that was prone to landslides, was gradually collapsing, and the villagers wanted to be moved elsewhere. The visit from the IFC was not prompted directly by the occurrence of landslides, however, but by the construction of an oil pipeline in the valley in which Dgvari was situated. The villagers feared that pipeline construction would intensify the frequency of landslides, and they looked to the pipeline company, which was led by BP, to address the problem. Geoscientific consultants, paid for by BP, had previously visited the village, taken measurements and produced a report, reaching the conclusion that although the villagers did need to move, the construction of the pipeline would not make the situation worse. A controversy therefore arose between the villagers and BP over whether or not the construction of the pipeline carried significant risks for the village, and whether the company had the responsibility for addressing the problem. It was this dispute that brought the IFC officials to the village of Dgvari.

In recent years geographers and social theorists have increasingly drawn attention to the critical part that materials play in political life. No longer can we think of material artefacts and physical systems such as pipes, houses, water and earth as the passive and stable foundation on which politics takes place; rather, it is argued, the unpredictable and lively behaviour of such

Material Politics: Disputes Along the Pipeline, First Edition. Andrew Barry.
© 2013 John Wiley & Sons, Ltd. Published 2013 by John Wiley & Sons, Ltd.

objects and environments should be understood as integral to the conduct of politics. Physical and biological processes and events, ranging from climate change and flooding to genetic modification and biodiversity loss, have come to animate political debate and foster passionate disputes. Yet if geographers have become interested in what has variously been described as the force, agency and liveliness of materials, thus probing the limits of social and political thought, then at the heart of this book lies an intriguing paradox: for just as we are beginning to attend to the activity of materials in political life, the existence of materials has become increasingly bound up with the production of information. Disputes such as those that occurred in Dgvari have come to revolve not around physical processes such as landslides – which have activity in themselves – in isolation, but around material objects and processes that are entangled in ever-growing quantities of information. The problem of the landslides of Dgvari was assessed by BP's consultants and Georgian geoscientists, as well as by the officials from the IFC, and the deteriorating condition of the villagers' houses was observed by numerous environmentalists and journalists over many years, as well as by myself. To understand the puzzling political significance of the landslides of Dgvari, I will suggest in what follows, we need to understand how their existence became bound up with a vast quantity of documents and reports that circulated between the village and the offices of ministries, scientists and environmentalists in Tbilisi, Washington, DC, London and elsewhere.

This book focuses on a series of disputes that arose along the length of the Baku-Tbilisi-Ceyhan (BTC) pipeline that now passes close by the village of Dgvari. In the period from 2003 to 2006 the BTC pipeline was one of the largest single construction projects in the world. Stretching 1760 km from south of Baku, the capital of Azerbaijan, on the Caspian Sea to the port of Ceyhan on the Turkish Mediterranean coast, it had first been conceived in the late 1990s when, in the aftermath of the break up of the Soviet Union and the first Gulf War (1990–91), international oil companies sought to gain access to off-shore oil reserves in the Caspian Sea, including the giant Azeri-Chirag-Guneshli (ACG) field. At the outset, the route of the pipeline through Georgia and Eastern Turkey was explicitly determined by geopolitical considerations, so as to enable oil exports from Azerbaijan to bypass alternative routes through southern Russia and Iran. Indeed, the pipeline was regarded from the late 1990s through the early 2000s as having enormous strategic importance both for the region and, according to some commentators at the time, for the energy security of the West. By 2004, the BTC pipeline employed nearly 22,000 people in Azerbaijan, Georgia and Turkey, with a projected cost of approximately $3.9 billion and the capacity to carry 1.2 million barrels of oil per day. While the pipeline was built by a consortium led by BP (BTC 2006), it involved a number of other international and national oil companies including the State Oil Company of Azerbaijan (SOCAR), Unocal, Statoil, Turkish Petroleum (TPAO), ENI, TotalFinaElf, Itochu and Delta Hess (see Table 1.1). It was also supported

Table 1.1   Institutions and organisations involved in the development and politics of the BTC pipeline

Participant oil Companies (equity stakes in 2003)	BP International and BP Corporation North America (30.1%); State Oil Company of Azerbaijan SOCAR (25%); Turkiye Petrolerri A.O. (TPAO) (6.53%); Statoil ASA (8.71%); TotalFinaElf (5.0%); Union Oil Company of California (Unocal) (8.9%); ITOCHU Corporation (3.4%); INPEX Corporation (2.5%); Delta Hess (2.36%); Agip (5.0%); Conoco Phillips (2.5%).
Contractors and consultants (selection)	Botaş (design, engineering , procurement, inspection); Spie Capag Petrofac (construction); WS Atkins (engineering consultants); Bechtel (engineering and procurement services); Environmental Resources Management (environmental and social impact assessment); Foley Hoag (human rights monitoring); Ernst and Young (sustainability monitoring); Mott Macdonald (lenders' environmental and social consultants); D'Appolonia S.p.A (lenders' independent environmental consultant); Worley Parsons (lenders' engineering consultant).
International financial institutions	International Finance Corporation – World Bank Group (IFC); European Bank for Reconstruction and Development (EBRD).
Commercial lenders (selection)	Royal Bank of Scotland (UK); Citigroup (US); ABN Amro (NL).
Export Credit Agencies	Eximbank (US); OPIC (US); COFACE (France); Hermes (Germany); JBIC NEXI (Japan); Export Credit Guarantee Department (UK).
International NGOs and related organisations	Amnesty International (UK); World Wildlife Fund for Nature; International Alert; Central and East European Bankwatch (CEE); Friends of the Earth (USA); Crude Accountability (USA). The Baku-Ceyhan Campaign: Friends of the Earth International; Kurdish Human Rights Project (KHRP); The Corner House (UK); Platform (UK); Bank Information Center (USA); Campagna per la Riforma della Banca Mondiale (Italy).
Regional NGOs (selection)	Open Society Institute (Azerbaijan and Georgia); Green Alternative (Georgia); Georgian Young Lawyers Association (GYLA); The Committee for Oil Industry Workers Rights Protection (Azerbaijan); Caucasus Environmental NGO Network (CENN); Association for the Protection of Landowners Rights (APLR) (Georgia); Centre for Civic Initiatives (Azerbaijan); Entrepreneurship Development Foundation (Azerbaijan); Institute of Peace and Democracy (Azerbaijan); Coalition of Azerbaijan Non-Governmental Organizations For Improving Transparency in the Extractives Industry.
NGOs involved in BTC Community Investment Programme (CIP) in Georgia	Care International in the Caucasus; Mercy Corps.

Sources: BTC/SRAP 2003a, BTC/PCIP 2003, BTC 2003b, 2006, Platform et al. 2003, House of Commons 2005b

by the US and UK governments, the International Finance Corporation (IFC)[1] and the European Bank for Reconstruction and Development (EBRD). Prior to its construction, the BTC pipeline had figured in the plot of the James Bond film, *The World is Not Enough*.

Yet the pipeline was much more than a vast financial and engineering project with security implications that stretched across three countries. For a period it was also viewed by many as a public experiment intended to demonstrate the value of a series of innovations in global governance that had developed progressively through the 1990s and 2000s, notably transparency, corporate social responsibility and 'global corporate citizenship' (Thompson 2005, 2012, Watts 2006, Lawrence 2009). Indeed, one of BP's explicit goals in developing BTC was to establish 'a *new model* for large-scale, extractive-industry investments by major, multinational enterprises in developing and transition countries' (BTC/CDAP 2007: 2, emphasis added, BTC 2003a: 7). It was, in particular, the first major test of the Equator Principles, the financial industry benchmark for 'determining, assessing and managing social and environmental risk' in project financing (Equator Principles 2003, Browne 2010: 172). This was a demonstration or test that would have to be performed in a region, the South Caucasus, in which none of the key parties – international oil corporations, investment banks, international NGOs – had much prior experience. In these circumstances, the parties involved in the development of BTC sought to carve out a space, simultaneously governmental, material and informational, within which this test could be performed and its results published. The BTC project is therefore remarkable not just because of its scale and complexity, or what was thought to be its geopolitical significance, but because an unprecedented quantity of information was made public about both the potential impact of its construction and how this impact would be managed and mitigated.[2] Indeed, as the project came to fruition in 2003, thousands of pages of documents about the pipeline were made public by BP, heading the consortium behind the project, while further reports were released by the IFC and other international institutions. At the same time, the pipeline attracted the attention of numerous documentary film-makers, artists, environmentalists, journalists, academics and human rights organisations.

The global oil industry has, of course, long been a knowledge production industry focused on the problem of how to locate and extract a complex organic substance that takes multiple forms from a range of distant and dispersed locations (Bowker 1994, Bridge and Wood 2005). Moreover, the oil industry has always been concerned with the problem of how to suppress, channel, contain or govern the potentially disruptive activity of materials and persons. In this light, the recent efforts to promote the virtues of transparency, public accountability and environmental and social responsibility have to be understood in the context of a longer history (Mitchell 2011). The story of BTC is in part a story of how the production and publication of information appears to offer capital a new, responsible and ethical way of managing the unruliness of persons and things. To understand the construction of the BTC pipeline, I suggest, we need to appreciate how its existence became bound up with the publication of information intended to effect its transparency. And to understand why and how its construction was disputed, we need to attend to the controversies that it animated, which did not just revolve around issues of geopolitics or the pipeline's relation to state interests, but also around quite specific technical matters concerning, for example, the likelihood of landslides, the impact of construction work on agricultural production, and the depth that the pipeline would need to be buried in the ground to protect it from sabotage. Indeed for a period, the BTC pipeline became the focus of an extraordinary range of particular disputes about what was known about its construction, its environmental impact, and even about the material qualities of the pipe itself.

I have already suggested that a case such as this poses a challenge to geography and social theory. The challenge is how to understand the role of materials in political life in a period when the existence of materials is becoming progressively more bound up with both the production and the circulation of information. At a time when social theorists and philosophers have drawn our attention to the agency, liveliness and unruly activity of materials, we need to be aware that the existence of materials is also routinely traced, mapped and regulated, whether this is in order to assess their quality, safety, purity, compatibility or environmental impact. This is not a new phenomenon; but the generation and circulation of information about materials and artefacts, including massive infrastructural assemblages such as oil pipelines, has come to play an increasingly visible part in political and economic life. One core argument of this book is that we need to develop accounts of the political geography of materials whose ongoing existence is associated with the production of information.

A second core argument follows. It responds to the claim that when information is made more transparent and publically available, rational and open forms of public debate should ensue (cf. Hood 2006). In this book I put forward an alternative account of the politics of transparency. I argue that the implementation of transparency, along with the growing salience of other core principles of transnational governance and social and environmental responsibility, foster new forms of dispute. The practice of transparency and corporate responsibility, I contend, does not necessarily lead to a reduction in the intensity of disagreement, although it does generate new concerns, sites and problems about which it matters to disagree. My central questions are geographical. In a period in which the virtues of transparency and environmental and social responsibility have been so insistently stressed, how and why do particular materials, events and sites become controversial? Why should quite specific features of the pipeline, such as its relation to the village of Dgvari, become matters of transnational political concern, while other candidate problems do not? If we understand the construction of the BTC pipeline as a demonstration of the practice of transparency, then, as we will see, the results of this vast public experiment turn out to be instructive.

The remainder of this introduction is organised into four parts. In the first, I introduce the idea of a public knowledge controversy, of which the case of the BTC pipeline is an example, and survey a number of key features of knowledge controversies in general, and public knowledge controversies in particular.[3] There is already a substantial literature on knowledge controversies, but here I introduce the concept of the political situation in order to highlight the way in which the spatiality, temporality and limits of any given controversy are themselves likely to be in question. I suggest that individual controversies, such as the dispute over the future of the village of Dgvari, are rarely isolated events. Rather, the relation between a particular controversy and other controversies and events elsewhere is likely to be uncertain and itself a matter of dispute. Individual knowledge controversies, I propose, need to be understood as elements of multiple political situations of which they form a part.

The second part of the introduction turns to the question of the way in which the properties, qualities and design of materials are bound up with the production of information. Human geographers have increasingly argued that they need to attend to what has variously been understood as the liveliness, agency and powers of materials as well as persons. I contend, however, that although this argument is an important one, it does not address the ways in which the existence and the activity of material artefacts have progressively been subject to monitoring, assessment, regulation and management. This observation has particular significance for the oil industry, which often operates in demanding environments in which the movement and activity of materials, including oil, land and pipes, may be difficult to manage and control. In this section I also highlight the critical importance of the production of social and political knowledge for the international oil industry when it operates in regions, such as the Caucasus, that are highly populated. Following Foucault's brief observation in the conclusion of the *Archaeology of Knowledge* about the need for analysis of the functioning of political knowledge, I suggest with reference to this study that such an analysis should include the social and political knowledge generated by, amongst others, BP, the international financial institutions and their critics.

In the third part of the introduction I return to consider the specificity of the politics of oil in the era of transparency, addressing two key issues. One concerns how the implementation of transparency raises questions about the range of matters in relation to which information is *not* made public. The other concerns how the length of the pipeline came to be constituted as a series of overlapping spaces of knowledge production and intervention – environmental, social, geoscientific, technical and legal – only some of which were rendered transparent. In the final section of the chapter, the argument turns back to consider the specific route of the BTC pipeline itself, pointing to the critical importance for the politics of the pipeline of the comparatively short section that ran through Georgia. The disputes that arose along the Georgian route became focal tests for the new model of corporate responsibility and transparency that was embodied in the construction of the pipeline.

Making Things Political

In 2005, the sociologist Bruno Latour and the artist Peter Weibel curated an exhibition at the Zentrum für Kunst und Medientechnologie in Karlsruhe, Germany, entitled *Making Things Public*. The exhibition built on the work carried out by historians and sociologists of science from the 1980s onwards on knowledge controversies (Collins 1981, Latour 1987). However, it placed this earlier work in an explicitly political frame. Conceived in the period immediately following the Iraq War of 2003, Latour took the infamous declaration made by US Secretary of State Colin Powell at the UN General Assembly that there were Weapons of Mass Destruction in Iraq as illustrative of the critical importance of both materials and knowledge claims about materials in public political life. The existence and the properties of objects, he contended, could generate passionate public disagreements. 'It's clear that each object – each issue – generates a different pattern of emotions and disruptions, of disagreements and agreements' (Latour 2005a: 15).

In *Making Things Public*, Latour therefore understood politics, in part, as a process in which objects can become the locus of public disagreement. In this view, objects should not be thought of as incidental to politics, but as integral to the disagreements and disputes that lie at the heart of political life. Here, I take Latour's account to be an expansion of the central claim made by theorists of radical democracy that at the heart of democratic politics lies a movement from antagonism to agonism. For Chantal Mouffe, '[t]o acknowledge the dimension of the political as the ever present possibility of antagonism requires coming to terms with the lack of a final ground and the undecidability that pervades every order' (Mouffe 2005a: 804, 2005b). In a democratic polity, Mouffe argues, dissensus should be the norm, not the exception. Nonetheless, the presence of antagonism can be addressed through the promotion of agonistic relations in which 'conflicting parties recognise the legitimacy of their opponents, although acknowledging there is no rational solution to their conflict' (Mouffe 2005b: 805). In these circumstances, decisions often have to be arrived at not by attaining a consensus, but in the face of persisting disagreement (cf. Waldron 1999a: 153–154). Studies of knowledge controversies take this perspective further, demonstrating that we should not expect that the disagreements that exist between experts will necessarily lead to a consensus (Stirling 2008), or that scientific evidence will provide a firm foundation or 'rational solution' on which political decisions can then be made. In a technological society, decisions are frequently made, and have to be made, in the face of persisting disagreement between experts about what the problems are, and how they should be addressed (Callon et al. 2001, Harremoës et al. 2002, Barry 2002).

But while radical democratic theorists point to the centrality of dissensus in political life, they say little about the existence and the importance of materials and objects, which frequently come to animate public knowledge controversies. Such controversies revolve around disagreements not just about the rights and interests of human actors and the identities of social groups (cf. Mouffe 2005b), but also about the causes of climate change, the safety of genetically modified organisms, the origins of diseases, the risks of floods and the consequences of nuclear accidents (Braun and Disch 2002, Callon et al. 2001, Kropp 2005, Demeritt 2006, Whatmore and Landström 2011). Through the emergence of new public knowledge controversies, the range of entities and problems that are taken to be the object of disputes continually shifts in time and across different settings. Sometimes public dispute may focus on the causes of the spread of a disease, on other occasions, on factors fuelling the decline in the population of a species. In the context of public knowledge controversies, some parties may seek to expand the range of sites in which disagreements can be both articulated and resolved far beyond the institutions of national government and parliamentary democracy to include farms, factories, research laboratories and the materials and bodies that they contain (Wynne 1996, Barry 2001, Hinchliffe 2001, 2007, Jasanoff 2006a, Law and Mol 2008). Rather than assume that public knowledge controversies are necessarily directed towards the institutions of the state, or that they must revolve around issues that are conventionally understood as political, the analyst of such controversies needs to attend to the historically and geographically contingent ways in which diverse events and materials come to be matters of public dispute. As we shall see, the very question of whether such controversies are framed as 'political' or not is commonly itself a vital element in the dynamics of the controversy. In this light, experts as well as non-experts can be viewed as minor political irritants (cf. Deleuze and Guattari 1987, Thorburn 2003), disrupting the certainties of what is conventionally understood to be the terrain of public debate by making visible problems and reanimating controversies that might otherwise be ignored or lie dormant.

A series of distinct areas of dispute come to the fore in public knowledge controversies, as earlier research indicates. These include the possibility of disagreement about evidence, including the quality, nature and relevance of the evidence germane to a controversy; about the procedures, techniques and instruments used to generate evidence that is considered relevant in a controversy; and about the theories that inform the production and interpretation of evidence (Collins 1981, Pickering 1981). Controversies may also come to centre on the nature of expertise, including the competence, qualifications, trustworthiness and interests of persons who claim to be experts or reliable witnesses (Shapin and Schaffer 1985, Collins and Pinch 1993); the manner in which, and degree to which, the work of experts is itself assessed or made publically accountable (Power 1997, Strathern 2004); the boundaries between what does and does not count as expertise; and the degree to which 'non-experts' may participate in the production of knowledge (Beck 1992, Epstein 1996, Wynne 1996). Previous studies of knowledge controversies have demonstrated that they may be especially difficult to resolve when, for example, no one body of experts can claim to possess a monopoly of expertise on a particular problem, or when the competence of experts is questioned. Moreover, non-experts may challenge the expertise of experts, or they may become involved in collaborative research with experts, thereby reconfiguring the boundaries of what counts as expertise (Callon et al. 2001, Barry, Born and Weszkalnys 2008). These are not new phenomena: the sources of expert authority have been disputed throughout the history of science, and it would be unwise to imagine that the distinction between experts and non-experts is less settled now than it once was (Schaffer 2005). As will become clear, all of these insights remain relevant if we are to understand the dynamics of the disputes that developed around the BTC pipeline.

But in addition, as we will see, the spatiality and the temporality of public knowledge controversies are themselves invariably in play and at issue. Knowledge controversies should not be understood as necessarily local, regional or global in their scope; nor are they exclusively immediate, medium-term or long-term in their significance. Indeed, the question of what should be included as part of any particular controversy, the significance of particular sites, its spatial dispersion and temporal extension, its history and future, its urgency: all these issues may become elements of the controversy. In a knowledge controversy, there may be disagreement about what the controversy is about; indeed disagreement can occur over whether there is a significant controversy, or just a minor technical problem that should be easy to resolve and need not occasion public debate at all. As should be evident from these varied elements, then, controversies have contested identities and multiple vectors of contention. In this sense the scope of a controversy 'is part of its effects, of the problem posed in the future it creates' (Stengers 2000: 67). It does not make sense to draw a hard distinction between local and global controversies for, as will become clear, the most apparently local processes can resonate in distant locations and at other times, and are sometimes intended to achieve exactly this effect. Nor are controversies invariably clearly bounded and distinctive in their scale, for one controversy can contain or portend another, manifesting nested or folded forms.[4] Controversies generally demand attention for a time, but the parties involved cannot be sure of, and do not necessarily agree about, either their significance or their resolution. Controversies may become intensely important for some participants, and not for others. What this amounts to is that none of the salient properties of a controversy – its scale and topology, its history and duration, its constituent elements, its privileged sites, its relevance for particular groups or classes, its shifting intensity and visibility, its identity or multiplicity – can be assumed (Strathern 1995). Controversies are neither static locations nor isolated occasions; they are sets of relations in motion, progressively actualised (May and Thrift 2001). They contain multiple sites and events that may lead to 'vast new chains of events' (Thrift 2006b: 549), interfering with the dynamic evolution of other controversies (Born 2010).

Thus it does not make sense to consider knowledge controversies as isolated and bounded events. Indeed, I want to propose another term – the political situation – that is intended to supplement the concept of knowledge controversy by pointing to and highlighting the indeterminacy, described in the previous paragraph, of what are the bounds and the significance of any singular controversy. While a particular controversy, such as the dispute surrounding the landslides of Dgvari, might appear to be self-contained, it may actually form only part of an ongoing and dispersed series of negotiations, debates or disputes. For many participants, those matters considered controversial in public, such as landslides in Dgvari, may be taken as more or less distorted signs of long-standing situations that are difficult to articulate but which, at least for those who are aware of their past significance, are recognised as somehow lying behind present events. Politicians, researchers and NGOs can also focus on specific controversial issues while recognising that this focus provides immediate tactical opportunities that make sense only in the context of broader strategies. In this way the specificity of the controversy may not matter that much. What matters more is the existence of a political situation, which may or may not be openly articulated, but in relation to which any particular controversy makes sense to participants and within which it is understood to be embedded. Indeed, one of the forms of expertise considered proper to the politician is the capacity to judge what is appropriate in the unfolding political situation so as to transform and reconfigure it in the midst of events (Crick 1962, Thrift 2006a). A good politician perceives when it makes sense to talk about the economy or crime, or specific economic or criminal events, knowing that the controversies that this is likely to provoke may produce advantages in a longer-term struggle for power. Likewise, as will become apparent in relation to the BTC pipeline, an environmental NGO may focus its attention on a very specific dispute around the pipeline not only or primarily because of the importance of the dispute in itself, but because of the relations that the NGO can seek to establish between a specific dispute and an evolving political situation – such as the irresponsible operations of transnational oil corporations. I refer to this process of establishing relations later as the logic of abduction (Chapter 4).

An analytics of the political situation therefore draws upon actors' own understandings not only of the significance of particular public controversies, but of their interrelations with other dynamics and events. For as I have suggested, actors often have an intuition or a conviction that a specific controversy forms only one element of a constellation of controversies. However, an analysis of the political situation cannot be restricted to actors' understandings of what is not made public in a controversy, or how it has been framed. Part of the value of the idea of the political situation is to point to the existence of dynamics that are not apparent in public but which are nonetheless critical to the evolution of the public knowledge controversy. The notion of the political situation therefore enables the analyst to establish connections between a series of different controversies that might otherwise seem unrelated. A political situation is not an underlying structure that governs the dynamics of a series of individual controversies. Rather, to call events a political situation is to argue for an expansion of the range of elements that should be considered when analysing a controversy, and to seek to analyse the sets of relations that are put in motion by any controversy.[5] The series of disputes that emerged along the BTC pipeline form part of a number of evolving political situations, as we shall see.[6] However, it follows from this stance that rather than treat the case of BTC as an example of a general phenomenon, I take the question of whether or not this case is of wider significance to be itself a matter of dispute (Barry 2012).

Up to this point I have focused on the ways in which knowledge claims can become matters of public disagreement. But if in certain circumstances knowledge and information are made public in order to politicise particular issues, it is equally necessary to recognise how in other circumstances knowledge and information can be made public in order to reduce, temper or moderate the level of passionate disagreement (Corbridge et al. 2005: 191). Making things public, in other words, is a strategy that can be employed either to politicise or to depoliticise a situation. A government, for example, may choose to delegate the role of producing economic forecasts to an office that is seen to be independent of the relevant minister partly in order to limit the degree to which such forecasts are contested. At the centre of this book is a discussion of the critical contemporary importance of transparency as a technique of governmentality, one that is often associated with ideas and practices of accountability. In particular, I focus on the way that the operation of transparency configures as 'public' certain objects and problems in the expectation that this will enable the form and intensity of public debate to be contained, by rendering it more rational and informed than it might otherwise have been (Strathern 2000, Best 2005, Held and Koenig-Archibugi 2005, Hood 2006, Jasanoff 2006b, Fisher 2010).

The escalating emphasis on transparency in transnational governance has occurred in synergy with the growth of the internet, which has made it possible to trace and map the course of public knowledge controversies online (e.g. Rogers 2004, Venturini 2011). Indeed, the internet has become a rich source of examples of public knowledge controversies spanning the entire range of the technosciences. But the increasing public availability of information and reports due to the growing stress on the importance of accountability and transparency also poses a significant challenge for those engaged in the study of knowledge controversies. For the public availability of information also raises questions about the processes by which certain types of information and analysis are made public while others are not, as well as about which actors and institutions play a more or less prominent part in fostering or inflaming public knowledge controversies. In an era in which the principles of transparency, public accountability and freedom of information have become global norms, the study of public knowledge controversies must address how these principles have become critical to the constitution and management of the boundaries between what is rendered public and what is not (cf. Shapin and Schaffer 1985: 342, McGoey 2007). It must interrogate the constitution of a field of public statements which is extraordinarily expansive, but which is equally characterised by certain marked limits and absences (Foucault 1972). The analysis of public knowledge controversies must therefore be concerned not only with the nature of information that is made public, but with what cannot become public information: it must be concerned, in other words, with information's constitutive outside (Butler 1993, Hayden 2010).

Governing Materials

Theorists of radical democracy have focused on the articulation of disputes between human collectives, the identities of which are shifting and relational. But as I have noted, they have had less to say about the importance of materials and technologies in political life and how the properties and behaviour of organic and inorganic materials – whether they are diseases, climate change, animal species, mineral resources or new technologies – themselves participate in such controversies. Central to the chapters that follow is the contention that material objects should not be thought of as the stable ground on which the instabilities generated by disputes between human actors are played out; rather, they should be understood as forming an integral element of evolving controversies. This occurs in part because the question of what the properties and behaviour of given materials are, or what they might become, can be focally what is in dispute. Knowledge of the properties of materials may not contribute to the resolution of a controversy, then, because that knowledge may itself be considered uncertain or controversial, while the competence of expert testimony about those properties may also be in question (Collins 1985). Indeed, controversy over the properties of materials is a critical feature of recent public knowledge disputes: consider the clashes that arose over the development of GM crops (Levidow 2001), the ongoing disputes over the safety of nuclear power and disposal of nuclear waste (Barthe 2006), and the disagreements that continually arise over the affordances of particular drugs (Fraser 2001).

At the same time, the complexities of natural and technological systems and the consequent difficulties involved in knowing and governing the behaviour of materials can provoke or contribute to the mutation of controversies. The activity and distribution of animals and biological materials may be hard to trace or to contain (Hinchliffe 2007, Hinchliffe and Bingham 2008, Law and Mol 2008); the slow transformation of metals, through creep or fatigue, is difficult to monitor (Barry 2002); the geological movements of the earth and its 'fearsome capacity' are hard to predict (Clark 2011). Yet all of these processes can have dramatic political consequences. One of the primary challenges facing those concerned with the analysis of public knowledge controversies is therefore how to rematerialise our understanding of politics (Braun 2008a, Bennett and Joyce 2010, Braun and Whatmore 2010). But this is not an easy task, on two counts. For one thing, it has to be achieved in a way that takes due notice of natural scientific accounts of the specificity of particular materials, while not assuming that this specificity should be understood only in natural scientific terms. But it must also be done in a manner that attends to the ways in which the properties of materials depend upon their changing relations with other material and immaterial entities. Understood in this way, the complex and shifting interrelations between material actors can have emergent effects, demanding what might be called an inter-object or inter-material analysis (Tarde 1969, Born 2010, Gregson and Crang 2010). Material artefacts never exist in isolation, but are themselves evolving entities that form part of a constellation of dynamic relations with other evolving entities (Pickering 1995, Barry 2005). As Whitehead observes, material entities are 'partially formed by the aspects of other events from their environments' (Whitehead 1985: 133).

It is ironic, however, that just as social theorists and philosophers are increasingly drawing our attention to the agency of materials, the properties and activity of materials have progressively become the objects of increasing levels of information production. The phenomenon is not new. As Simon Schaffer reminds us, the practice of the assay once made a vital contribution to the formation of a connection between evidence and political authority. Assays established a measure of control over the properties and qualities of quotidian substances, from tobacco to cloth and drugs, while at the same time 'extraordinary provisions were made to discipline and assess through strict examination and central policing not merely the trade in these commodities but the qualities of analysts themselves' (Schaffer 2005: 306). Efforts to regulate the unruly properties of materials have therefore long been entwined with attempts to produce disciplined and reliable forms of expertise.

In the present day, as I have said, materials are subject to a growing range of modes of information production, such that their existence is bound up with the production of information. Two broad reasons for this development can be discerned. First, in the context both of the escalating costs of raw materials and energy, and of the demands of consumers and industry, there is an abiding emphasis on assessing the *performance* of materials: 'classification, measuring, modelling, testing and adjusting materials is a constant process' (Harvey and Knox 2010: 137, see also Bowker and Star 1999, Lloyd Thomas 2010). This observation applies to the oil industry like any other. Indeed, as we shall see, a notable controversy emerged precisely around the question of whether an innovative material component of the BTC pipeline had been adequately tested and its performance properly evaluated (Chapter 7). Second, material assemblages are the object of a growing range of *regulatory* requirements governing such issues as environmental waste, biosecurity, safety and energy use (Bulkeley and Watson 2007, Gregson et al. 2013). The production of information about materials is therefore intimately associated with the growth of national and transnational regulatory zones, regimes that govern, measure and monitor the impact of materials on both persons and the physical environment (Barry 2001, 2006, Dunn 2004, Fisher 2008).

In an earlier article I highlighted the ways in which the generation and transformation of an increasing range of material entities is bound up with the production of information (Barry 2005). Consider, for example, the idea of a 'proven oil reserve', which refers to the quantity of oil that is technically feasible and economically profitable to produce from a particular field (Ahlbrandt 2006). A proven oil reserve is not a representation of an existing object – a reservoir of oil simply waiting to be exploited; nor is it an inventory of a given stock of material. Rather, a proven oil reserve is a virtuality: it is a quantity of oil that might be extracted economically, using available technology and given prevailing market prices. It is a virtual entity that condenses a potential relation between a material object and known technical processes and economic conditions. In effect, estimates of proven oil reserves serve as projective devices or protentions: they enable companies, governments, managers, engineers and investors to extend their understanding into the future, by envisaging certain actions, without necessarily ever knowing precisely what exists in the present (Born 2003, 2007). In this case, the production of information (about an 'oil reserve') translates a complex set of materials into a new object of economic calculation. Or consider the problem of how to detect and prevent pipeline corrosion, which, if unchecked, may lead to catastrophic failures. On the one hand, the problem of corrosion has driven the development of a series of devices, such as cathodic protection, and materials, such as epoxy coatings, that have progressively transformed the material components of pipelines themselves. On the other hand, pipelines are now subject to a variety of forms of monitoring through, for example, ultrasound techniques and measurements of electrical potential or magnetic flux. These measurements do not represent corrosion as a complex electrochemical process, but rather point to the existence of defects that may require repair. In this way measurement and monitoring add to the existence of a pipeline, encasing it in an array of figures, traces and samples that may enable the potentially disruptive effects of corrosion to be contained. Disputes about the environmental impact of a pipeline, such as BTC, do not revolve around an isolated material object. Rather they engage with a material object whose integrity is formed and progressively transformed through multiple layers of information production (cf. Barry 2005, Lloyd Thomas 2010).

In the oil industry the production of scientific information has particular characteristics. For some of the forms of natural scientific and engineering expertise deployed in the oil industry are best understood as field sciences: they analyse problems in particular settings, distant from the laboratory (Schaffer 2003, Livingston 2003, Powell 2007). In the field, the engineer or scientist does not encounter materials in a pure form, but in relation to the specificity of their changing environment: corroding in the soil or sea, impacted by dust, vibration or landslides, monitored by technicians or robots, neglected by managers or workers, or subject to tapping or sabotage (Bowker 1994, Kennedy 1993, Selley 1998). It follows that a technology such as pipeline, tanker or drill that may work reliably in one location may not work so well in another, where prevailing environmental conditions are different, appropriate management systems do not exist, or there is a lack of expertise or equipment to ensure that the performance of instruments is regularly checked (cf. Graham and Thrift 2007). In these circumstances, the engineer or scientist must be alert to the complexity of his or her relations with the dynamic environment of which s/he is a part and to whose existence s/he also contributes. Oil companies have therefore long been concerned with the technical and managerial problems, as well as the financial costs, of containing and monitoring the unruly properties of both materials and persons. But this has to be achieved not in the carefully controlled conditions of the scientific laboratory, but in the open-ended and potentially unstable environment of the field (Chapters 2, 6 and 7).

However the complexities do not end here. For oil companies and financial institutions not only draw on natural scientific expertise; in the context of growing demands for assessments of the impact of the oil industry on the economy, society and human rights, they increasingly draw also on the expertise of social researchers (Clark and Hebb 2005, Watts 2005, Bridge 2008). Yet exactly how it is possible to make the infrastructure of the oil industry into a field site for social research is far from clear (cf. Livingstone 2003: 48). How can fieldwork provide the kinds of reliable knowledge required in order to demonstrate that the oil company is either succeeding in addressing, or failing to address, its growing list of social responsibilities? Through what processes can experts, non-experts and (indeed) counter-experts generate evidence about such critical matters as social impacts, human rights abuses, corruption or violence? And to what extent does social research on the social and economic impact of the oil industry take into account the impact of information that has already been made public about social and economic impacts (cf. Luhmann 2002: 219, Esposito 2011)? How can the assessment of impact assess the impact of assessment itself?

The importance of social research on the oil industry and its impacts poses a challenge for analysts of knowledge controversies. For it means that participants in knowledge controversies may not only dispute what are typically regarded as 'scientific' matters – such as the nature of evidence, the competence of experts or the reliability of instruments; but they may also question what are generally regarded as 'political' matters, such as the interests of the public, the organisation of public debate, inequalities in power and resources, or the relation between experts and democracy (Callon et al. 2001, Jasanoff 2006a, Fischer 2009, Asdal 2008). During the period of my fieldwork along the BTC pipeline, such disputes had particular intensity. In public controversies of this kind, actors disagree about the identities, opinions and interests of affected publics, as well as the nature of the settings in which debate should occur. They disagree also about the unacknowledged financial interests and political affiliations of other participants in the controversy. From the point of view of the philosophy of science and the sociology of scientific knowledge, it is completely unsurprising that actors have different views about such matters and dispute the views of others. After all, 'the specialist in political science deals with a dimension of human societies that is not the material for an "objective" definition, practiced in "the name of science", because in itself this dimension corresponds to an invention of definitions' (Stengers 2000: 59).

But from the perspective of those concerned with the study of politics, disputes about such issues as what politics is, what the political opinions of others are, and what interests have not been made public ought to matter a great deal. In this book I focus not only on disputes over matters of scientific fact, but on the ways in which actors' own political knowledge, political theories, political practices and political analyses figure in the dynamics of disputes. In this respect the approach taken here has a certain resemblance to the work of those political anthropologists who start out from actors' situated experiences and accounts of politics and the state (Gupta 1995, Humphrey 2002, Navaro-Yashin 2002, Ssorin-Chaikov 2003, Tsing 2005, Lazar 2008, Hibou 2011, Reeves 2011, see also Jeffrey 2013). At the same time, I take a bearing from Foucault's suggestion near the end of the *Archaeology of Knowledge* that it is possible to give an account of how political knowledge is 'inscribed, from the outset, in the field of different practices in which it finds its specificity, its functions and its network of dependencies' (Foucault 1972: 194). His later analysis of governmentality arguably falls short of this ambition, in as much as it has led to accounts of political rationality and governmental technique that tend to be decoupled from a sense of the contingency and the passion of political life (Navaro-Yashin 2002, Thrift 2006a, Walters 2012). In contrast, in the chapters that ensue, I analyse the operation of techniques of global governmentality such as transparency that have developed 'from the outset' in the midst of controversy.

An Experiment in Transparency and Responsibility

The BTC pipeline, I have suggested, was a public experiment: it was intended to be a demonstration of the value of transparency in the oil industry. In this way it was expected to provide a new model for large-scale extractive industry investment that would, in effect, respond to the growing recognition of the damaging consequences of the industry's operations for the environment, economy, society and human rights (Watts 2005). Indeed, by the late 1990s it was increasingly thought that the promotion of greater transparency was part of the solution to the 'resource curse' that was said to afflict the economies of many oil producing countries. This recognition led to the announcement of a new Extractive Industries Transparency Initiative (EITI) by Tony Blair at the World Summit for Sustainable Development in Johannesburg in 2002 (Chapter 3). But in a period in which transparency has been elevated to a general principle of global governance, a crucial consequence tends to be overlooked: it is, as I pointed out earlier, that the question of what is and what is not made public about the operations of the oil industry has become a matter of dispute (cf. Rancière 1998). Transparency tends to be understood as a normative principle or demand that can be justified or criticised on extremely general grounds. But in fact, as I argue in this book, the very exercise of transparency informs the way that certain processes and events, such as environmental and social impacts, become the objects of public dispute, while others do not.

A common response to the apparent opacity of the oil economy to external scrutiny follows on. Rather than stress the value of transparency as a more or less formalised process, this response involves attempts to reveal some of the oil industry's secrets, thereby uncovering its lack of transparency in practice (cf. Urry 2003: 116–117). This is the approach taken in a series of more or less well-researched books, articles and documentary films that detail, *inter alia*, the deals made by oil companies with governments, the links between oil, corruption and violence, and what are claimed to be the real environmental impacts of oil industry operations. It is a response that both establishes and relies upon connections between investigative journalism, insider knowledge and leaks, academic research and the work of NGOs who are more or less critical of the industry (e.g. Rowell et al. 2005, Leech 2006, Ghazvinian 2007, Peel 2009, Bower 2010, Maas 2009, Muttit 2011, Bergin 2011,

Marriott and Minio-Paluello 2012). Critics engaged in this kind of investigation question not only the claims made by experts working for the oil industry, but their authority and legitimacy as experts. Dissident industry insiders and journalists track the deals made between governments and companies. Environmental and human rights NGOs support their critical arguments through evidence that contradicts the public claims made by companies and their consultants (Chapter 2). Former oil company geologists dismiss industry projections of future oil production (Deffeyes 2001), only for their views to be dismissed by others (Clarke 2007). In short, the expertise of industry specialists is contested by a growing range of counter-experts and 'lay protests' (Beck 1992: 162). In relation to the oil economy, then, the substantial growth in the quantity of information made public, partly in response to the demand for transparency, has been met by an escalation of public disputes about the value and the significance of this information. Questions about what is made public and what is not, and about what is kept secret or confidential and what is not, have themselves become vital political issues (Barry 2006, Neyland 2007). In effect, the operations of transparency create something of a catalytic surface on which new antagonisms can both form and be resolved, fomenting a particular type of counter-politics, one that does not aim to undermine the principle of transparency but, on the contrary, demands greater transparency and the availability of ever more information. In this very real sense, critics of the oil industry have generally sought not to interrogate the principle of transparency, but rather to expand, shift and deepen the realm of its operation.

It follows that if, for many of its proponents, the principle of transparency is expected to have global applicability, in practice the principle is applied only in particular spaces and to specific objects. In other words, the production of information about the activities of the oil industry renders certain objects, materials, problems and spaces visible, while others are not. The Extractive Industries Transparency Initiative, for example, was initially concerned with quite specific pieces of financial information and operates primarily at the level of the nation-state (Chapter 3). In comparison, the BTC project gave the idea of transparency a much more ambitious interpretation, as well as a distinct and novel spatial form. Rather than restrict the operation of the principle of transparency to financial information, it extended it to a host of other matters, including environmental and social impacts. Moreover, as the chapters that follow show, in pursuing transparency, BTC sought to forge multiple spaces of information production along the route of the pipeline – environmental, geological, technical, financial, legal and social – spaces that were not necessarily isomorphic with one another and that sometimes overlapped, effecting a kind of palimpsest (Chapter 5). At the same time, the knowledge claims constituting these spaces were networked through the offices of BTC and its contractors, consultants and NGOs in London, Baku, Tbilisi and elsewhere, as well as banks and international organisations based in the UK, the United States and other oil consuming countries (Barry 2006, Bridge and Wood 2005, Bridge 2009, Mitchell 2011, Vitalis 2009: 266, Bridge and Le Billon 2012: 62–65). If, as I argued earlier, the existence of materials is increasingly bound up with the production, circulation and publication of information, the construction and operation of the BTC pipeline depended on the production of information not only about the pipe itself, but about the project's impacts.

In this way, by rendering certain aspects of the impact of its operations visible, BTC also attempted to mark out – however provisionally – the limits of its social and environmental responsibilities. Nonetheless, the boundaries between the interior and the exterior of the spaces mapped and made visible through the production of information were not fixed. After all, some of these spaces had only a temporary existence during the period of construction, and could be amended as the project progressed (Chapter 5). Moreover, given the pipeline's existence as a socio-material assemblage, the boundaries of the 'impact' of the pipeline could be transformed unexpectedly due to accident, neglect, sabotage or the occurrence of natural disaster; or previously unrecognised 'impacts' could gradually become apparent (Chapter 6). The unpredictable and ungovernable behaviour of materials and persons could therefore intrude, all too obviously, on the integrity of these spaces. The constitution of such governable spaces (Watts 2004) depended not only on the work of BP and its lawyers and consultants, but on the activities of regulators and lenders who monitored its operations, and those of critics who contested the company's published accounts of its impacts by producing alternative accounts that drew attention to further sources of pollution, damage, violence or injustice (Tables 1.1 and 1.2). In these circumstances, the boundaries manifesting the spatial limits of transparency became ambiguous, shifting and disputed.

The Georgian Route

The BTC pipeline stretches across three countries. It begins in Azerbaijan at the Sangachal terminal south of Baku, taking oil extracted from the giant Azeri-Chirag-Guneshli (ACG) field in the Caspian Sea and transporting it westwards, crossing the Georgian border near the town of Gardabani. Its passes close to the industrial city of Rustavi before turning south of Tbilisi, the Georgian capital, near the military base of Krtsanisi. Its route subsequently traverses the southern half of Georgia, passing through a region in which there is a substantial Armenian minority, before skirting the resort of Bakuriani and the borders of the Borjomi-Kharagauli national park, running along one side of the valley in which the village of Dgvari is located. Finally it crosses the Turkish frontier near the town of Akhaltsike, entering a region in which there is a mixed Turkish and Kurdish population, before looping southwards and westwards past the eastern Turkish city of Erzurum,

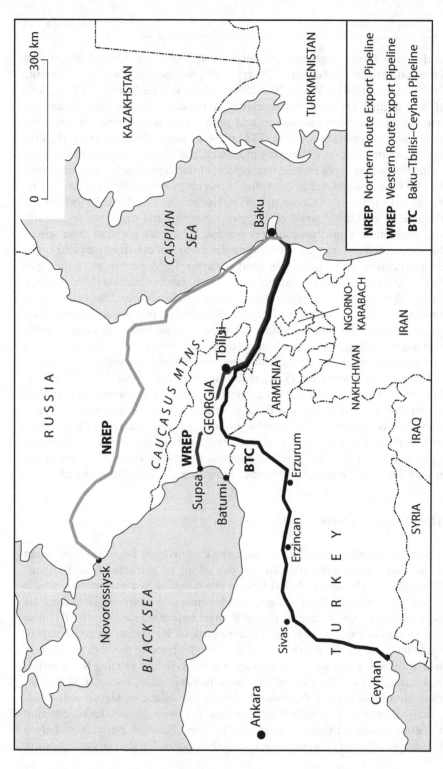

Figure 1.1 Route of the BTC pipeline. Map prepared by Ailsa Allen, School of Geography and the Environment. Reproduced by permission of the University of Oxford

Table 1.2   Timeline of the BTC pipeline project

1991	March. Independence of Georgia from the Soviet Union
1993	October. Heydar Aliev becomes President of Azerbaijan
1994	September. BP, Statoil and Amoco sign contract with Azerbaijan government to develop ACG oil field in the Caspian Sea
1995	November. Eduard Shevardnadze becomes President of Georgia
1999	April. Baku-Supsa pipeline opens
	November. Intergovernmental Agreement BTC signed by Azerbaijan, Georgia and Turkey (Chapter 2)
2000	October. BTC Host Government Agreements signed (Chapter 2)
2001	Environmental and Social Consultation begins (Chapter 5)
2002	May. Draft Environmental and Social Impact Assessment (Chapter 5)
	June. Beginning of 60 day public consultation period in Georgia (Chapter 5)
	June. International NGOs begin fact-finding missions, until 2005 (Chapter 2)
	August. BTC pipeline company formed
	October. Tony Blair announces Extractive Industries Transparency Initiative (EITI) at the World Summit for Sustainable Development in Johannesburg (Chapter 3)
	December. Georgian government agrees environmental permit (Chapter 2).
2003	March. Demonstration outside of BP offices, Finsbury Circus, London
	April. BTC construction starts
	June. Final Environmental and Social Impact Assessment made public (Chapter 5).
	June–October. IFIs public consultation period (Chapter 5)
	August–September. IFIs hold Multi-stakeholder forum meetings in Borjomi and Tbilisi as well as in Azerbaijan and Turkey (Chapter 5)
	September. Geoscientists carry out Dgvari landslide study (Chapter 6)
	September. Demonstration outside the offices of the EBRD, London (Chapter 5)
	November. Rose Revolution in Tbilisi (Chapter 2)
	November. IFC and EBRD agree to finance BTC project (Chapter 2)
	November. Problems with cracks in SPC 2888 in eastern Georgia (Chapter 7)
2004	Spring–Summer. Blockages in villages across Georgia (Chapter 6)
	June. Nino Kirtadze films in the villages of Tadzrisi and Sakire (Chapter 8)
	July. IFC ombudsman investigates complaints in Dgvari, Sagrasheni and other villages (Chapters 6 and 8)
	July. Georgian government halts construction work in the Borjomi region (Chapter 2)
	September. BBC4 screen documentary on BTC (Chapter 6)
	December–January. House of Commons Trade and Industry select committee inquiry (Chapter 7)
2005	May–October. Azerbaijan and Georgian sections of BTC pipeline inaugurated
	August. BTC engineering consultant visits Sagrasheni, Atskuri and other villages (Chapter 6)
2006	June. First oil delivered by BTC pipeline to tanker in Ceyhan, Turkey
2008	August. Georgian-Russian War
	Reports of bombing of BTC pipeline near to Akhali Samgori (Chapter 9)
2009	February. Azerbaijan becomes the first EITI compliant country (Chapter 3)

Sources: Host Government Agreement (2000a, 2000b, 2000c), BTC/ESIA (2003), BTC (2006, 2009), EITI (2012)

and finally towards the Turkish Mediterranean coast and the terminal at the port of Ceyhan. In total 433 km of the route passes through Azerbaijan, 248 km through Georgia, and 1070 km through Turkey.

While the length of the Georgian stretch of the pipeline is comparatively short, it nonetheless became the focus for some of the most intense disputes along the route of the pipeline. There are a series of contingent reasons as to why this might have been the case, which I discuss further in Chapter 2. One reflects the particular significance of the pipeline in Georgian politics during the last years of the government of Eduard Shevardnadze. Although the Georgian government was only expected to receive approximately $50 m in annual transit fees following the completion of pipeline construction (Billmeier et al. 2004: 8), and relatively few Georgians would be employed in the long term by the BTC company, Shevardnadze had stressed the vital importance of the BTC pipeline for the country, the economy of which had been devastated following the collapse of the Soviet Union and the disastrous war that followed. However, the proposal to allow the route of the pipeline to pass through the environmentally sensitive region of Borjomi and Bakuriani was reported to have been initially opposed by the Georgian Environment Minister, Nino Chkhobadze, less than a year before Shevardnadze's own departure from government following the Rose Revolution of November 2003 (Table 1.2). I return to consider the critical significance of the Borjomi region to the politics of the pipeline in more detail in Chapter 2.

Another reason why Georgia became the location for a series of serious disputes stems from both the state of civil society and the relatively open climate for political action. Georgian civil society was strongly supported by Western governments and NGOs in the early 2000s, while Georgian NGOs had good networks of contacts with their Western counterparts (Hamilton 2004). The radical Georgian environmental NGO Green Alternative, for example, which became involved in the disputes surrounding the village of Dgvari, received support from Oxfam and was also a member of the influential Prague-based Central and Eastern European Bankwatch network. Moreover, although the Shevardnadze regime is said to have been characterised by widespread corruption (Kukhianidze 2009, Schueth 2012), there was arguably a substantial degree of freedom in Tbilisi in the period prior to the Rose Revolution (L. Mitchell 2009: 39, Wheatley 2005). In these conditions it was possible for the wisdom of the president's decision to allow the route of the BTC pipeline to pass near Borjomi to be discussed and debated in the Georgian media. By comparison, there had been little Western support for civil society organisations in Baku, while opposition to the Azerbaijan government had a record of being heavily managed or suppressed (Cheterian 2010). While I do not discuss the development of BTC in Azerbaijan in any detail, I return in Chapter 3 to consider the broader efforts to promote transparency in Azerbaijan through the Extractive

Industries Transparency Initiative (EITI). In Turkey international NGOs, including Amnesty International, raised questions about the relation between the construction of the pipeline and the protection and abuse of human rights, while also criticising the terms of the Host Government Agreement between the Turkish government and BTC (Amnesty International 2003, Baku-Ceyhan Campaign 2003a).

In these circumstances, in the chapters that follow I focus primarily on the constitution and contestation of the stretch of the BTC pipeline that ran through Georgia. This is instructive because, as I have said, a series of intense disputes emerged along the Georgian stretch during the period of land acquisition and pipeline construction that lasted from 2002 to 2005, some of which came to the attention of the IFC and the European Bank for Reconstruction Development, as well as a range of international NGOs, researchers and activists. If we take BTC to be a demonstration of the practice of transparency and global corporate citizenship, then the route of the pipeline through Georgia, in particular, provides some of the clearest evidence of the outcome of this public experiment. In addition, as a state that plays a critical role in oil transportation but not in oil production, Georgia makes an intriguing location for a study in the politics of oil. In this respect the focus of this book complements the growing body of social research on the politics of oil producing states or regions (Coronil 1997, Sawyer 2004, Watts 2004, 2006, Soares de Oliveira 2007, Valdivia 2008, Reed 2009, Overland et al. 2010, Behrends et al. 2011, Yakovleva 2011, McNeish and Logan 2012, Rogers 2012). Indeed, while the development of both the ACG field and the BTC pipeline has been vital to the political economy of Azerbaijan since the mid-2000s (Kalyuzhnova 2008, Lussac 2010a&b, Overland et al. 2010, Cornell 2011), the direct economic significance of the BTC pipeline to Georgia is quite limited. To reiterate, although I focus on the politics of the Georgian route, my argument is not that the economy and politics of Georgia in recent years have been dominated by oil. Rather, I give an account of why specific materials, objects and places along the route of the pipeline became objects of both local and transnational dispute.

The Archive

At the heart of the analysis that follows is an archive. It consists of a huge body of documentation made available by BP, along with public reports by the International Finance Corporation and the European Bank for Reconstruction and Development, both of which gave financial support to the BTC project. The archive includes, *inter alia*, the agreements made between the oil company and the three host governments, Azerbaijan, Georgia and Turkey, assessments of environmental impacts, details of compensation rates for losses in agricultural production, the procedures for and results of consultations with affected communities, maps of projected routes, archaeological surveys, oil spill plans, sites of river crossings, and reports from the various bodies established to monitor the project. The publication of this archive online, running to many thousands of pages, had complex implications that will become evident in later chapters.[7] But on encountering this archive it is immediately apparent that it has a double function and is aimed at two broad readerships. On the one hand, the archive can be understood as a projective and managerial device, documenting the environmental and social reality of a region, setting down how the company intended to intervene and the commitments that it entered into, as well as assessing its present and past performance. During the period from 2002 to 2006, this aspect of the archive was frequently updated as the construction project progressed. On the other hand, the publication of the archive was intended to meet the demands for transparency, accountability and corporate social responsibility made by international financial institutions, investors and civil society organisations. The dynamic interaction between these two functions of the archive plays a critical part in the narrative that follows.

In itself, the archive contains a remarkable body of documents. It tells us a great deal about how an oil company claims to know the world in which it intervenes, and how it intended to intervene in that world on the basis of the knowledge that it had generated (cf. Burton 2005, Stoler 2009). Indeed elements of the archive were intended to be performative: to bring into being the very reality to which they referred. But in the main period of the project the archive also had, and still has, clear and systemic limits. It does not tell its readers much, for example, about the relations between BP, its partners, and their contractors and consultants. It contains little account of the relations between the international oil companies, on the one hand, and the national governments and national oil companies, on the other. It does not inform readers, except in the most general way, about the politics and political economy of the region. And if transparency is intended to reduce corruption and violence, then the existence of corruption and violence, and their scope, complexity and effects, are addressed only in the margins of documents. In short, the archive embodies the principles of transparency and corporate social responsibility but, in the very same instant, it remains resonantly silent about some of the 'key problems that the enactment of these same principles are intended to manage and address. In effect, by drawing a sharp division between what is considered to lie inside the realm of good governance and transparency and what lies outside this realm, the constitution of the archive consistently evades what Béatrice Hibou has termed 'the political problem' (Hibou 2011: 282). The archive amounts to an extraordinarily rich and prolix source, while also being marked by 'a limited set of presences' (Foucault 1972: 119, see also Ahiska 2010).[8]

The archive was and remains a projective device, one that narrates the future of a project while at the same time reconstructing and reflecting on its past.

But its documents also contain a set of claims and promises towards which critics of the BTC project directed their fire. The archive was faulted for its factual errors and its absences, as well as for the discrepancies between what the pipeline company promised it would do in the documentation and what it actually did in practice. Indeed, the generation of the archive was mirrored by the formation and circulation of a number of much smaller counter-archives that documented these criticisms, much of which appeared online (cf. Tarde 2001 [1890]). The archive was, then, much more than a description of a project. Its constitution and contents became integral to the disputes that emerged in relation to the pipeline, as well as their resolution. In the chapters that follow I argue that we have to understand the significance of the archive in the midst of events – that is, in relation to ongoing claims, interventions, interpretations and decisions in which it played a part. The generation of the archive was not only an instrument of management and a manifestation of the practice of good governance; it also had the effect both of channelling disagreement towards particular sets of objects and problems, and of acting as a catalyst for a series of intense disputes about matters of fact.

In describing the controversies within which the pipeline figured, I nonetheless argue for the need to trace the limits of what is made public in the archive. In doing so my aim is not to uncover hidden causes behind public statements; nor is it to reveal what has been covered up or displaced through the overproduction of information; nor is it to demonstrate that the archive is in fact a fiction, and that transparency is merely a façade. Instead, I interrogate how practices of making things public and of criticising what has been made public have come to be central to the governance and politics of oil. In this light it is vital to explore the boundaries of what is contained in the archive not in order to expose the scandal of what has been kept secret, but rather because the question of what has and what has not been made public became integral to an array of disputes surrounding the construction of the pipeline. The creation of the archive is a remarkable achievement; yet at the same time, a number of my informants were aware of the limits of what had been published. A careful scrutiny of both the archive and the counter-archives directs us towards the significance of dynamics and events, such as strikes and village blockages, which were only ever addressed in the margins of published documents.

The analysis that follows derives from five sources. One is the archive of documentation about BTC that I have just described, which was generated around the work of BP, their consultants and partners, and which remains accessible on the BP Caspian website. The second is a series of other reports published both by the international financial institutions and by international and local NGOs, many of which assess the reports produced and commissioned by the BTC company, in this way informing the development of the archive and adding to it a further layer of commentary, as well as eliciting further responses. The third source consists of the frequent press reports attracted by the development of the pipeline, as well as at least three documentary films and several other artistic and photographic projects that took the pipeline as their object. I draw particularly on the work of the Georgian documentary film-maker Nino Kirtadze and the Czech film-makers, Martin Maraček and Martin Skalsky, as well as contemporary British and Georgian news reports. The fourth source informing this book is a body of interviews with over one hundred participants in the events described, including officials working for governments and international organisations, oil company managers, engineers and community liaison officers, and professionals engaged in corporate community investment programmes, along with consultants, journalists, activists and academic scientists and social scientists. Approximately half of these interviews were carried out in Tbilisi and rural Georgia, while others took place in Baku, Ankara, Kars, Sarikamiş, Oslo, Prague, London and Washington, DC. Most of the interviews were conducted in 2003–4, in the period of pipeline construction, while others took place in September 2010 when I returned to Georgia during the period of the pipeline's operation. The research benefits enormously from the contribution of a research team that included, at different times, Meltem Ahiska (in Turkey), Farideh Heyat (in Azerbaijan), and Joanna Ewart-James and Alex Scrivener (in the UK and Georgia). In line with the professional ethics of academic social research, I have attempted to anonymise my interviewees and interlocutors throughout this book.

The analysis draws, finally, on a series of field visits that I made in 2004 and 2010, sometimes accompanied by BTC company community liaison officers and sometimes by local NGOs critical of the project, to villages along the pipeline route. Sociologists and anthropologists of scientific knowledge have often sought to witness the production of information at first hand. In a context in which the process of information production was itself highly politicised, this was not possible. Nonetheless, fieldwork made it possible both to trace some of the experiences of consultants, officials, reporters and activists who had already come to the same locations, and to be alert to the ways in which the visits of outsiders had had consequences not just internationally, but also for those more immediately affected by the pipeline's development. In drawing on this material I do not make a sharp division between my own fieldwork practice and the practice of company advisors, consultants and activists. In order to produce information, they too relied on some of the same documents, in conjunction with interviews and brief periods of fieldwork. The similarities in our practices render the differences more distinct (Riles 2006: 89).

The research for this book occurred during a period in which the relations between some international NGOs and the oil company were sour, indeed antagonistic.[9] Fieldwork was carried out in the midst of controversy. Yet in my practice of participation observation, rather than being embedded in one organisation or aligned with one position, I moved back and forth across the lines, attempting not to be partisan, tracing the course of disputes from as many directions as possible. In doing so I was guided both by Georgina Born's approach to multi-perspectival ethnography, and by Marilyn Strathern's idea that ethnography is an open-ended, non-linear method of data collection, such that '[r]ather than devising research protocols that will purify the data in advance of analysis, the anthropologist embarks on [an] exercise which yields materials for which analytical protocols are often devised after the fact' (Strathern 2004: 5). On some occasions my lack of explicit affiliation may have aroused suspicion about my own motivations and identifications. This was quite unsurprising in the circumstances. But in practice, it was possible to avoid becoming directly implicated in events as they happened and to maintain a position that could scarcely be described as external to those events, or merely disinterested, but which nonetheless minimised its own immediate effects. In a world in which the publication of information, or the possibility of its publication, could create intense feedback, my maintenance of a public silence was not a threat. At the same time, at least some of those involved in the events that I describe here were able to distance themselves from the public positions of the organisations for whom they worked and to address the complexity of the situation as they experienced it, as well as the limitations of their own understanding and knowledge. I remain extraordinarily grateful to these people for their reflexive insights and for their generosity in sharing them with me. The study that follows was not intended to intervene in a series of unfolding disputes, but to contribute to their rethinking at a moment when critical reflection would be possible. If this study gives voice to a particular view, my aim is that it should be informed by the insights of those who, immersed in events, were aware of their complexity, but were not in a position to articulate this awareness at the time.

Overview of the Chapters

At the heart of this book is a geographical puzzle. How is it possible to understand why particular materials and sites along the route of the pipeline came to be of transnational political significance, while others did not? The answer to this question should involve, as I have argued, an account of the operation of transparency along the pipeline; but it should also entail an analysis of the politics of materials. My explanation of why individual disputes occurred will necessarily, however, be limited. My account is inescapably partial. I highlight the importance of certain dynamics, while only indicating the significance of a multitude of others. Following Foucault's injunction my aim is to multiply causes, while acknowledging that there will, inevitably, be more to be said (cf. Foucault 2002a).

Over the course of the next three chapters I introduce four ways in which particular locations and materials acquired transnational political importance. The first two of these revolve around the relation between the construction of the pipeline and Georgia itself. A first reason why the pipeline came to be of transnational interest derived from the claim, made by many Western commentators, that the pipeline had critical strategic importance as a route for the transportation of oil in the wake of the break-up of the Soviet Union. A second way, in contrast, was powerfully informed by Georgian domestic politics but also had considerable international significance: it concerned the putative impact of the pipeline on the sensitive environment of the Lesser Caucasus. It was in this context that the western section of the Georgian pipeline route, including the village of Dgvari, acquired heightened political salience. Thirdly, specific problems and locations – such as traffic vibration, or the location of beehives and trees along the pipeline route – came to be controversial and to attract considerable transnational interest because of the project's espousal of transparency. Finally, specific materials and locations attracted transnational interest because of how they could be made to demonstrate that the oil corporation was, or was not, acting ethically. In all these ways, and for all these reasons, a very specific set of materials and locations came to figure as possible – or what we might call candidate – elements of transnational political situations.

Chapter 2 does not begin with a discussion of BTC directly, but draws a contrast between three ways in which the relation between Georgia and the politics of oil have been figured. Taking a cue from Bertrand Russell's claim that it was the presence of oil that led to the Soviet invasion of Georgia in 1921, the chapter begins with an interrogation of the manner in which Georgia came to be understood as an 'energy corridor' from the Caspian Sea to the West. This geopolitical analysis of the importance of Georgia is contrasted with a radically different way of framing the politics of oil: one that was focused not on the strategic calculations of national governments, but on the physical geography of Georgia itself. I highlight, in particular, the critical importance of the mountainous region of south-west Georgia to the domestic politics of the pipeline. Finally, I turn to consider how the transnational governance of the BTC project became controversial. Indeed, according to some NGOs the development of the pipeline completely failed to conform to a series of relevant international guidelines and standards.

Chapter 3 probes in greater detail the centrality of both the idea and the practice of transparency to the contemporary politics and governance of oil. The focus of this chapter is not on BTC itself, but on the operation of a more modest experiment in transparency, which developed in around the same period: the Extractive Industries Transparency Initiative (EITI). Drawing on recent studies in the anthropology and history of economic expertise, the chapter interrogates the design of transparency as a technique for governing the range of matters about which it is possible to disagree. The chapter dwells on the case of Azerbaijan, stressing the importance of witnesses to the operation of transparency and the difficulty of assembling a body of witnesses that are able to judge the value and the accuracy of the information that transparency generates. It highlights the differences between the limited and controlled experiment in transparency associated with EITI and the more radical, extensive and conflictual experiment in transparency, involving the eruption of serial controversies that occurred along the length of the BTC pipeline.

Chapter 4 turns from a consideration of the importance of transparency in the politics of the pipeline to the significance of ethics as it became both embodied in a range of international agreements and principles and expressed through the practice of corporate social and environmental responsibility. The chapter develops three arguments. One concerns the way that the ethical conduct of corporations, including their commitment to principles of social and environmental responsibility, has come to be both demonstrated and assessed. A second argument centres on the importance of particular material artefacts, accidents and events in the ethicalisation and politicisation of oil. Here the analysis highlights the importance of what C.S. Peirce termed 'abduction' to the politics of oil, examining how specific materials and issues can be made both to encapsulate and to transform a political situation. Finally, the chapter considers the manner in which the ethical conduct of oil corporations has become the object of political research. The argument focuses, in particular, on the inventive and influential practice of Platform, a London-based group of artists and researchers who both traced and revealed a series of problems along the pipeline route.

If Chapters 2 to 4 are concerned with the ways in which the route of the BTC pipeline came in general to be politicised, drawing out a series of political vectors and dynamics that figure to different degrees in the evolving political situations around the pipeline, Chapters 5 to 8 attend directly to a series of very specific disputes that emerged around the pipeline's development, construction and operation. Running through the analyses in these chapters are three ongoing concerns, highlighted both in this introduction and in the chapters. The first centres on the idea of BTC not only as a material artefact, but as a socio-material assemblage. A second concern is with the importance of the mobilisation of the public to the practice of transparency. The third concern is to recount how particular material entities, including buildings, lorries, land and pipes, acquired the remarkable political salience that they did.

Given that a vast quantity of information about the BTC pipeline was made public, various manifestations of the 'public' were expected to be ready to be consulted about the project, and to become interested in this information. Chapter 5 interrogates the ways in which the problem of how to assemble relations between diverse publics and the pipeline was both addressed and contested. In particular, the chapter examines how the construction of the pipeline was bound up with the formation of a narrow corridor of land along the route within which the population were defined as 'pipeline affected communities'. Disputes over the construction of the pipeline developed both around the question of what information should be provided to diverse publics, including 'the affected communities', and whether such communities had been properly consulted.

Prior to the construction of the pipeline, the oil company had commissioned an assessment of the pipeline's environmental and social impact along its entire length. However, during the course of the project, the impact of construction work became the focus of a spate of controversies at a number of points along the route. Chapter 6 probes these controversies, analysing in particular the dynamics of those disputes that turned on whether or not damage to the social or environmental infrastructure did or did not constitute 'impacts', and, consequently, whether the company could or could not be held responsible for them. Dwelling on the cases of two Georgian villages, Dgvari and Sagrasheni, the politics of both of which escalated transnationally, it shows how disagreements over the difference between the space within which 'affected populations' were located and the evolving and uncertain space of potential 'impacts' played a critical part in the emergence of disputes along the pipeline.

If Chapter 6 addresses the importance of material processes that were the potential source of 'impacts', Chapter 7 probes how the materiality of the pipeline itself acquired transnational political significance. It develops further the larger question of the politics of materials, as well as the multiplicity of the pipeline as an 'informed material'. Empirically, the chapter focuses on how the problem of assigning responsibility for the poor performance of a physical component of the pipeline – a coating material called SPC 2888 which covers joints between sections of the pipe – came to be debated extensively in the British House of Commons. Theoretically, it interrogates the relation between the properties of materials and the contingency of politics.

The application of the principle of transparency to the BTC pipeline project, as I have explained, led to the generation of a vast and evolving archive of material, including assessments of environmental and social impacts. Chapter 8 centres on an interrogation of this archive, pursuing how the transparency of the archive itself transformed the objects and processes that it described. Focusing on the economic interventions of the oil company, the chapter contrasts the transnational visibility of compensation, community investment and the 'affected population' with the transnational invisibility of a range of issues including the politics of labour. The contrast highlights how the enactment of transparency intensifies the significance of particular processes, generating feedback, while obscuring others. In Chapter 9, the Conclusions, I return to a number of core themes from the book including the politics of transparency and its limits and the relation between knowledge controversies and political situations.

Chapter Two
The Georgian Route: Between Political and Physical Geography

In the wake of the break-up of the Soviet Union, the location of Georgia in the South Caucasus appeared to give it a new strategic significance for the West. In this view, the importance of Georgia did not reside in the fertility of its soil or its mineral resources, but in what it enabled others to avoid. As many analysts observed, the position of the country meant that, in principle, it would be possible to export Caspian oil to global markets while avoiding alternative routes through Iran, Armenia or Russia. However, in the 1990s, analysts had little concern with the precise route that Caspian oil exports would take across Georgia; nor were they interested in the political and economic history of the country, nor its physical geography. Rather, they drew attention to the value of Georgia as a route for oil transportation or, in other words, as 'an energy corridor' (BTC/RR 2003:27).

From the early 2000s onwards, however, the interest of political analysts in the Georgian route declined. By this time, the idea that a pipeline should be built from the Caspian to the Turkish Mediterranean coast had been agreed between a consortium of oil companies, led by BP, and the governments of Azerbaijan, Georgia and Turkey. In place of the earlier preoccupation with the strategic location of Georgia, a radically different series of concerns emerged. From around 2002, the route of the proposed pipeline through Georgia came to be the focus for a vast multi-disciplinary research project, carried out on behalf of the BTC company. During this period, potential pipeline routes through Georgia were mapped in diverse ways and at increasing levels of detail by experts in the geosciences, engineering,

Material Politics: Disputes Along the Pipeline, First Edition. Andrew Barry.
© 2013 John Wiley & Sons, Ltd. Published 2013 by John Wiley & Sons, Ltd.

biodiversity, archaeology and social and environmental impact assessment. The route of the pipeline was subsequently explored by a series of national and international NGOs, journalists, film-makers and artists. In this way, the constitution of the Georgian route became increasingly bound up with the production of information about engineering, environment, politics and society. In this second phase, the materiality of the pipeline and its immediate environment came to matter in a way that had been unanticipated by the analysts who had drawn attention to the strategic value of the Georgian route in the late 1990s.

The chapter is in four parts. In the first part I recall how the discussions in the 1990s were prefigured in the writings of Bertrand Russell and Leon Trotsky in the 1920s. However, my interest in Russell is not just because his writings anticipate the preoccupations of more recent researchers with the significance of the Georgian route. Rather, I turn to Russell because of the striking contrast in his writings between his reductive geopolitical assessment of the strategic importance of Georgia and the analysis of the socialist system that he derived from his own travels across the Soviet Union. Russell's visit to Russia directs us towards the significance of the practice of what I term political fieldwork.

The second part of the chapter concerns the way in which the area of the Caspian-Caucasus was analysed in the mid-1990s as a region in which the principal political actors are taken to be states that possess a strategic interest in the control of the production and transportation of oil. Here I trace how in the period after the collapse of the Soviet Union, Georgia came to be figured once again as an export route for Caspian oil. In this specific context, Western analysts showed no interest in the everyday life of the population of the country, their ethnic diversity or their historical experience of foreign intervention, nor were they concerned with the physical environment within which the pipeline would have to be constructed. Rather, they positioned Georgia in a geopolitical grid such that its strategic value for the West was self-evident. The third part of the chapter turns to the critical importance of the physical geography of Georgia to the politics of the pipeline in Georgia itself. Where Western analysts had shown little interest in physical geography in the 1990s, the mountains of the Lesser Caucasus powerfully informed the domestic politics of the pipeline in Georgia in the 2000s. I focus in this section on the controversy that erupted around the question of the precise route that the pipeline should take through the mountains of south-west Georgia in the region of the spa town of Borjomi. In this light, the case of the BTC pipeline should force us to return to and rethink the proposal, first made in 1887 by the British geographer Halford Mackinder, that those concerned with the study of politics must necessarily be concerned also with the study of the physical environment (Mackinder 1887: 216). As readers of Mackinder might have anticipated, the disputes that began to emerge along the route of the pipeline through Georgia revolved, amongst other things, around questions of physical geography.

The final part of the chapter focuses on the emergence of a new transnational politics that developed progressively through the 1990s, which focused on the conduct of social and environmental research, good governance, transparency and corporate social responsibility (Rajak 2011, Thompson 2012). By the early 2000s, Western European activists, journalists and artists – interested in problems of environmental damage, human rights and social injustice – came to visit the region in growing numbers. Like Russell, they sought to interrogate the formation of an emerging economic and political system not only through its documents, but also through brief periods of fieldwork and direct observation of real conditions. Informed by these explorations of 'corporate colonialism' (Marriott and Muttit 2006), transnational public debate came to focus less on the actions and interests of states, as it had previously, and turned towards a series of quite specific 'impacts' of the oil industry on society, human rights and the environment. If, in the 1990s, the route of the BTC pipeline through Georgia had been explicable in a larger geopolitical and strategic context, by the mid-2000s the gaze of researchers was directed towards an apparently more limited territory running along the projected 1760 km pipeline, including the 248 km Georgian section that ran from Gardabani in the east to Vale near the Turkish border.

Political Fieldwork

In *The Prospects of Industrial Civilization* (1923), Bertrand Russell criticised Trotsky's attempt to justify the incorporation of Georgia into the Soviet Union. For Trotsky, 'wherever the fiction of self-determination becomes, in the hands of the bourgeoisie, a weapon directed against the proletarian revolution (as in the case of Georgia), we have no occasion to treat this fiction differently from the other "democratic principles" perverted by Capitalism' (quoted in Russell 1923: 86). Russell responded that while he did not entirely disagree with the 'theoretical attitude' informing Trotsky's argument, what actually lay behind Soviet actions in Georgia was, although Trotsky did not say so, the control of oil. Russell proceeded to put forward the proposition that 'it is contrary to all sound socialist doctrine that the oil should be the private property of Georgia, but there is no better reason why it should belong to Soviet Russia. It ought to belong to a world-wide combination, which would ration it to the various according to their needs and their economic suitability for using it' (Russell 1923: 86–87, see also Ryan 1988: 100). In effect, Russell questioned Trotsky's own commitment, and the commitment of many socialists, to internationalism: 'if England and the United States were socialistic, they would have no better right to the [Panama and Suez] canals than they have now, but might be just as anxious to retain them as Trotsky is to retain the Georgian oil. National socialism therefore will not solve our problem' (Russell 1923: 87). Russell did not address Trotsky's counter-claim that British interests in Georgia were also about oil: 'there are even now White Guard organizations under the high-sounding title of "Liberation Committees" (a title that does not prevent them from receiving money subsidies from British and Russian oil magnates, Italian manganese magnates, etc.)' (Trotsky 1975 [1922]: 98).

Russell's disagreement with Trotsky's account of the justification for the Soviet invasion of Georgia in 1921 raises two key issues. The first concerns the history of Georgia as a territory of shifting geopolitical consequence to major foreign powers. In an essay first published in 1904, Halford Mackinder famously stressed the vital significance of what he called the 'pivot of history', which he located in Central Asia: 'is not the pivot region of the world's politics that vast area of Euro-Asia which is inaccessible to ships, but in antiquity lay open to the horse-riding nomads, and is today about to be covered by a network of railways?' (Mackinder 1904: 434, see also O'Hara 2005, Kearns 2009). Mackinder had earlier claimed that the new science of geography was concerned precisely with the question of the interaction of man in society and his environment. As a result, as he had argued previously, 'no *rational* political geography can exist which is not built upon and subsequent to physical geography' (Mackinder 1887: 214, emphasis in original). By implication, Britain's imperial strategy should be grounded upon the rational analyses of geographers. In Mackinder's analysis, Georgia and the Caucasus lay on the border of the strategic pivot area (Mackinder 1904: 435), and his essay gave Lord Curzon, the British Foreign Secretary in 1919, 'what was believed to be a "scientific" basis for foreign policy' in the region (Livingstone 1992: 195).[1] Subsequent writers have affirmed the significance of Mackinder's intervention; Dodds and Sidaway, for instance, proclaim that 'the development of geopolitical texts about power, space and international relations in the twentieth century frequently drew upon Mackinder's legacy' (Dodds and Sidaway 2004: 293, see also O'Tuathail 1996, O'Tuathail et al. 1998, Kearns 2004, 2009).

Yet while the geopolitical value of Georgia might be understood in relation to the continuing resonance of Mackinder's analysis, for Russell and Trotsky the significance of Georgia was to be found in its relation to the location of natural resources. Russell made a claim that is increasingly common today: that the foreign policies of powerful nation-states are often driven by a desire to gain access to and control over oil supplies. In Russell's eyes, Georgia was of interest to the new Soviet government because of the wealth of its resources. In this respect he was mistaken. Contrary to Russell's claim, in fact Georgia had little oil of its own. As one contemporary source noted, oil production from the Tiflis (Tbilisi) fields in 1914 amounted only to 581 tons compared to nearly 7 million tons from the Baku fields and 1.9 million tons from the other major field in the region, near Grozny (Ghambashidze 1919: 63).[2] As Trotsky rightly argued, Georgia's most strategically significant natural resource at this time was not oil but manganese, a vital element in the production of steel. Indeed, in the period prior to the First World War, Georgian mines were said to be the source of approximately 25 per cent of total global manganese production (ibid.: 164). Nonetheless, since the late nineteenth century, oil had come to play a part in its political economy. As early as 1883, a railway line had been built between Baku, the centre of the oil industry in Azerbaijan on the Caspian Sea, and the Georgian Black Sea port of Batumi (Kautsky 1921, Chapter 4, Akiner 2004). The port was occupied by British troops in 1919, an action condemned by Trotsky and the Georgian Soviet, who noted that 'during the occupation ... the Georgian Menshevik's policy towards Soviet Russia was especially insolent and provocative' (Congress of the Georgian Soviets 1975 [1922]: 113). In relation to the international oil economy, the importance of Georgia even in this period was not due to its resources, as Russell imagined, but to its constitution as a route.[3]

The second issue raised by the British philosopher is less obvious. Russell's book *The Prospects of Industrial Civilization* was written following a visit to Russia in 1920 that he made with a delegation sent by the Trades Union Congress, with the aim of learning directly about the political and economic conditions of the country under a Bolshevik government (Monk 1996: 574, Ryan 1988). During his visit Russell briefly encountered Trotsky at the Opera House in Moscow (Russell 1920: 34) but, in the circumstances, was only able to have a 'banal' conversation with him. Following a subsequent interview with Lenin, in which Lenin dismissed the possibility of a non-violent revolution in England, waving 'aside the suggestion as fantastic' (ibid.: 38), Russell spent several weeks with the delegation observing Russian life outside Moscow and Petrograd and travelling along the Volga by boat as far as Astrakhan, near the northern shore of the Caspian (Monk 1996: 579–580). Although he reasoned that Bolshevism was attractive to a country 'in distress', the philosopher was unimpressed both by Trotsky and Lenin and by the practice of Bolshevism, for 'every kind of liberty is banned for being "bourgeois"; but it remains a fact that intelligence languishes when thought is not free' (ibid.: 151).

Russell's thoughts on industrial civilisation and Bolshevism highlight how the delegation's travels through Russia can be understood as a means of knowledge production. Historical and sociological studies of knowledge production have tended to focus on the work of natural and social scientists, often associated with recognisable disciplines or fields (e.g. Latour and Woolgar 1986, Collins 1985, Pickering 1995). Although Russell's research was not intended as a contribution to a particular discipline, and he had no expertise in social research, he nonetheless conducted a type of fieldwork that drew together his reading of Marxist texts with direct observations of political and working conditions in Russia and of daily life in Moscow. His research was not systematic at all, but it nonetheless amounted to what

Michel Foucault, following his visits to Tehran in 1978, would later call the 'reporting of ideas' (Osborne 1999, Foucault 2001). A contrast can be drawn between Russell's broad condemnation of Soviet policy towards Georgia and his diagnosis of everyday life in post-revolutionary Russia. The first mode of political analysis was based on an assessment of the growing centrality of oil to foreign policy; the second was based, however, on what might be called political fieldwork (cf. Driver 2001: 199). Russell's disagreement with Trotsky was about more than geopolitics or even matters of political principle; it was a dispute about the real conditions emerging under socialist rule, as observed by the British philosopher. Trotsky did not respond to Russell, but he did draw attention to the limitations of any conclusions based on excessively rapid observation, criticising Kautsky for defending the Menshevik government in Georgia after Kautsky's admission that he 'did not see anything except what could be seen from the window of the train or in Tbilisi' (Trotsky 1975: 12).[4]

Geopolitics, Oil, the Caucasus

Russell's interest in the justification for the Soviet invasion in Georgia in 1921 resonates with the growing Western interest in Georgia in the 1990s. After the demise of the Soviet Union, political analysts in the US and Europe became preoccupied, like Russell, with the possibility of what they termed 'transition' from one political and economic system to another (cf. Humphrey 2002). Moreover, for many recent commentators, the parallels between the late nineteenth to early twentieth centuries and the late twentieth to early twenty-first centuries were obvious enough (Kleveman 2003, O'Hara 2005). The arrival of international oil companies on the Caspian Sea in the mid-1990s echoed the period from 1880 when Western companies played a vital role in the development of the Baku oil fields (Yergin 1991: 57–63, Bowker 1994) and the presence of British troops in Tbilisi and Batumi during the first years of the Menshevik government seemed to parallel events in the 2000s, when the US government started to provide training for the Georgian army through an extensive programme of 'democracy assistance' in Georgia (L. Mitchell 2009, see also T. Mitchell 2011). For many political analysts in the late 1990s, the strategic significance of the Caspian and the Caucasus stemmed precisely, as it had for Russell, from the location of oil fields. More generally, the Caspian-Caucasus – as it was sometimes called (Gökay 1999) – had become an object of strategic calculation: 'a power vacuum was created [with the demise of the Soviet Union] with lines of control less certain. Notwithstanding lesser hegemonic control, there has been no corresponding abatement of interest [in the region] … old and new players are engaged in various interest seeking games, under different banners with different agendas' (Amirahmadi 2000a: vii). Like

Mackinder, the imperial geographer, Western analysts conceived of the strategic value of Georgia primarily in terms of its importance for Western energy security. And like Russell, the liberal philosopher, many of them reasoned that oil lay at the heart of its geopolitical significance.

The apparent continuities between the writings of Russell and recent political analysts suggest the appropriateness of one form of critical response. This is that all of these accounts of the Caucasus should be viewed as elements of a broader historical system of geopolitical representation. Drawing on the writings of Foucault, Antonio Gramsci and Edward Saïd, amongst others, writers in the tradition of critical geopolitics have interrogated forms of geopolitical reasoning and the constitution of geopolitical imaginaries, whether they are to be found in the work of strategists and academic commentators, such as Mackinder, in statements of foreign policy, or in fictional and popular texts, including film, television and literature (O'Tuathail 1996, Dodds 2007, Macdonald et al. 2010, Power and Campbell 2010). For these writers, 'geopolitics is ... not a centred but a decentred set of practices with elitist and popular forms of expression' (O'Tuathail and Dalby 1998: 4).

Certainly the Caucasus and surrounding regions have played a particularly salient part in the geographical imagination of Euro-American geopolitics (cf. Gregory 2004). In English, this system of representation not only includes works of political and economic analysis, international relations and political geography, but also, as critical geopolitical analysts would lead us to expect, fictional and autobiographical narratives. Notable examples includes John Buchan's novel, *Greenmantle*, in which the Eastern Anatolian city of Erzurum comes to be a strategic location during the First World War (Buchan 1993 [1916]: 186), and Fitzroy MacLean's autobiographical *Eastern Approaches*, which describes the author's clandestine explorations to Baku and Samarkand in the 1930s while working in the British Embassy in Moscow (MacLean 1949). In the film *The World is Not Enough*, James Bond becomes embroiled in the intrigue surrounding the development of an oil pipeline through the Caucasus, a region that had become dangerously destabilised by the collapse of the Soviet Union. The film, according to Klaus Dodds, 'draws on mainstream political concerns within Russian, Turkish and Euro-American security debates even if it completely subverts the great power realities of the post-Cold War era' (Dodds 2003: 148). As these diverse sources suggest, this strategically vital territory does not have a specific name, but is a border zone, lying roughly between Russia, Turkey, Iran, the Black Sea and Afghanistan. This is an imaginary region of espionage, political instability, corruption, violence and ethnic conflict, considered critical both to the security of the British, Ottoman and Russian Empires in the late nineteenth and early twentieth centuries, and to the energy and military security of the US, the UK and Russia at the end of the twentieth century. The contours and borders of the region are not given, but both inform and are redefined by the area's various conflicts.

An interpretation of the 'Caspian' and the 'Caucasus' as figures of Western geopolitical discourse can, however, only be a starting point for an analysis of the manner in which the Georgian route became an object of transnational politics. Indeed, an analysis based solely on geopolitical discourse is limited and has two main drawbacks. Firstly, it fails to address how observations and reports by experts in both the natural and social sciences could become the focus for disagreement, both transnationally and locally. In this respect, critical geopolitical analysis has not sufficiently addressed the importance of expertise in international politics including, in particular, the significance of fieldwork (cf. Megoran 2006, Jeffrey 2013: 4).[5] Secondly, the problem that political analysts confronted in the 1990s was always more than a matter of representation. Their assessments of the politics of the Caucasus, however well or poorly informed, were intended to interpret the ways in which states and international oil companies should act in the near future, in relation to an emerging series of opportunities, obstacles, threats and risks. Such assessments were both anticipatory and performative: pointing to the possibility and the challenges of interventions to come (Thrift 2000: 381, 2008, Anderson 2007, Barry 2010, Toal and Dahlman 2011). The value of such analyses depended not so much on their exhaustiveness or accuracy, but on the degree to which they could inform or justify action in the present. Assessments were situated in the midst of events and, in principle, depended on what we might call the experts' situated political knowledge of the shifting intentions and interests of states.

According to many analysts of state politics and international relations, the strategic significance of the region in the 1990s came down to its energy resources. The actions of the 'major players' – nation-states and international oil companies – were therefore explicable in terms of their desire to gain control over the region's oil: 'the Caspian region developed in the course of the last decade of the 20th century into an important strategic and security arena with an apparent potential to provide additional energy for the world economy' (Ehteshami 2004: 63). The US government, in particular, had fostered a special interest in the Caspian-Caucasus, reckoning that it was vital to its efforts to reduce the dependence of the US on Middle East oil supplies: 'not infrequently, the Caspian Sea zone is described as the world's new great energy frontier and a place of great importance in general and to US interests in particular' (Ebel and Menon 2000: 1). According to William Engdahl, US interest in Georgia continued to be all about oil even, or especially, after the Rose Revolution of November 2003, which led to the presidency of Mikheil Saakashvili: 'Georgia, lying on a key pipeline route from the Caspian to Ceyhan in Turkey, was a *de facto* US protectorate by the beginning of 2004' (Engdahl 2004: 265).[6]

Western political experts from the early 1990s onwards developed more or less reductive or subtle analyses of the intersecting, conflicting and shifting interests and motivations of different states. Of course, it was acknowledged that the newly independent nation-states of the Caspian had an interest in modernising their ageing oil and gas industries in order to generate export revenues. This was particularly true of Azerbaijan following the disastrous war over Nagorno Karabakh (Dragadze 2000: 131–157, Hoffman 2000, Bradshaw and Swain 2004). Moreover, although Soviet geologists had previously identified major offshore oil fields in the Caspian, these could not be developed without the technology and expertise of Western companies. At the same time, the Caspian offered international oil companies access to reserves that were outside of OPEC, and were also not directly threatened by the conflict and political instability in the Middle East, particularly following the invasion of Kuwait by Iraq in 1990. Potentially at least, the Caspian-Caucasus appeared to provide part of the solution to what was and is widely assumed to be the problem of US 'energy security', while helping to sustain the dominant position of US and British companies in the international oil industry. Indeed, the initial agreement between the government of Azerbaijan and a consortium of oil companies, led by BP and Amoco, which opened up the prospect of access to the giant offshore Azeri-Chirag-Guneshli (ACG) field was named 'the contract of the century'. Although initial estimates of the scale of Caspian oil reserves were over optimistic, the Caspian-Caucasus offered what David Harvey (2006) would call a 'spatial fix' for the international oil industry, enabling it to secure access to reserves following a period in which it had lost control of the oil fields of the Middle East (Bromley 1991).

Given that the Caspian-Caucasus was thought to have such strategic importance, the literature in the latter half of the 1990s weighed up the political costs, benefits and, above all, the risks of different strategies. Such calculations involved a series of distinct but related elements, which are given different emphases by different commentators and analysts. In particular, the legal status of the Caspian following the break-up of the Soviet Union was unresolved (Granmayeh 2004, Gizzatov 2004, Bayulgen 2009). This was not just a question of the conflict between the Caspian states over the control of natural resources; it also involved the difficulty of 'translating geographic description and classification into a normative legal practice' (O'Lear 2005: 172). The political and economic value of Caspian energy resources therefore depended on their scientific and legal mediation (cf. Powell 2010). At the same time, the Caucasus was marked by a series of conflicts that could be fomented by major powers, including not just the 'frozen' conflict between Azerbaijan and Armenia over Nagorno Karabakh (de Waal 2004, Kaldor 2007), but also those surrounding the disputed status of the separatist regions of Abkhazia and South Ossetia in Georgia. After all, in the early 1990s 'Russia [had] exploited Azerbaijan's and

Georgia's conflicts and internal weaknesses to pressure them to fall into line on joint defence of the external borders of the former Soviet Union, the maintenance of Russian bases on their territory and the deployment of exclusively Russian ... peace-keepers for the region's conflicts' (Herzig 1999: 49). With the increasing involvement of the Turkish government in discussions over possible export routes for Caspian oil, the conflict between the Turkish state and the PKK in eastern Turkey also had to figure in analysts' assessments of political risk.

Framed in this way, much of the public and academic debate in the late 1990s in the West focused on the question of the specific route that major pipelines might have to take (Forsythe 1996, Roberts 1996, 2004, Adams 1999, Ebel and Menon 2000, Amirahmadi 2000b, Rasizade 2002, Kellogg 2003, Ehteshami 2004, cf. Bouzarovski and Koneiczny 2010). An existing pipeline, constructed during the Soviet period, ran from the northern shore of the Caspian through Chechnya; another possible route through Russia avoided Chechnya. In this light, the Chechnyan wars of 1994–96 and 1999 had to be understood as 'far from being a foreign policy adventure ... [and more as] part of a concerted effort to keep the Caspian's outbound supply routes within the Russian federation' (Ehteshami 2004: 66). The Russian routes, ending at the Black Sea port of Novorossiysk, had the advantage that the terrain made pipeline construction relatively easy in comparison to farther south; but they had the disadvantage from the point of view of US strategists, as well as the governments of the newly independent states of the South Caucasus, that they passed through Russia. Moreover, 'one of the principal purposes of any energy export system is to safeguard the political independence secured by energy producers of the Caspian through the attainment of a degree of economic independence' (Roberts 1996: 8–9, see also Ebel and Menon 2000: 5, Nourzhanov 2006).

Furthermore, the construction of pipelines to the Black Sea would increase tanker traffic through the already congested Bosphurus, unless further pipelines were built from, for example, Odessa in the Ukraine, or Constanza in Romania, to connect the Black Sea to either Baltic or Mediterranean ports, while bypassing Istanbul (Köksal 2004: 165). A southerly route, through Iran, was apparently supported for a time by BP, the major international oil company investing in Azerbaijan (BTC/ESIA 2002i: 3-7), but US opposition to Iran ruled out this possibility (Browne 2010: 163). In addition, there were three potential 'central' routes through either Armenia or Georgia, all of which avoided both Iran and Russia (Croissant 1999, Jeter 2005). While it was technically feasible to build a pipeline through Armenia, this possibility would depend on the resolution of the conflict between Azerbaijan and Armenia over Nagorno Karabakh, which was unlikely to occur in the near future (Herzig 1999, Kaldor 2007). One route through Georgia would parallel the existing railroad bringing Caspian oil to Supsa or Batumi on the Black Sea coast. This became the route for so-called 'early oil' from the Caspian, although its use would also lead to increased tanker traffic through the Bosphurus. A final potential route would bring a pipeline through Georgia and then through north-east Turkey, bypassing the major areas of conflict between the Turkish state and the Kurdistan Workers' Party (PKK) in the south-east. In this nexus of state interests and strategies, logistical bottlenecks, frozen conflicts and security considerations, the position of Georgia became critical. From the perspective of Western strategists, the disadvantage was that Georgia had been politically unstable in the recent past and contained regions – Abkhazia and South Ossetia – that had broken away from government control following the civil war of 1991–93. However, with the return of Eduard Shevardnadze to Georgia in 1992, and his accession to the presidency, the problem of political stability seemed to be partially solved, as the new president appeared able to unite different factions of Georgian politics, while accepting a limited Russian military presence in the country (Suny 1994: 331, Wheatley 2009: 124). Moreover, the Turkish government had developed close relations with Azerbaijan in 1991–92, a period when the Turkish government had pursued a pan-Turkic foreign policy across the Caspian and Central Asian republics of the former Soviet Union (Köksal 2004). However, while this policy had collapsed as early as 1992 (Robins 2002: 284), from 1995 onwards the Turkish government argued for the possibility of constructing a pipeline through Georgia. Despite initial scepticism from the Azerbaijan International Oil Company (AIOC),[7] the US government supported this option. According to the US government, the pipeline was certainly a commercial proposition, but it was also in the 'strategic interest of Azerbaijan, Georgia and Turkey' (Jones 2001), by establishing a vital cord between Turkey, Georgia, Azerbaijan, and the US (Roberts 2004: 83–84). In this way Georgia, and the South Caucasus more generally, came to be figured as what BTC company's own review of region called a potential 'energy corridor' (BTC/RR 2003: 27).

Timothy Mitchell has argued that the distinction between the 'state' and the 'economy' has to be understood as an artefact of a particular political technique. This distinction, he suggests, should not be understood 'as a boundary between two distinct entities but as a line drawn internally, within the network of institutional mechanisms through which the social and political order is maintained' (Mitchell 1999: 77). Mitchell's argument is directly relevant to this case. For in the strategic political analyses of the decision to construct a pipeline through Georgia that I have described, the economic calculations of the oil companies – and the series of assumptions they made about, for example, construction costs, land compensation payments, transit fees, the size of Caspian oil reserves, future oil price levels, and environmental, political and security risks – are largely invisible. In effect, political analysts during the 1990s drew a demarcation between the domain of state politics and interests and the sphere of economic calculation, even though BP and other corporations certainly had an interest in the economic viability of the Georgian route in this period, however much public deliberation around it was explicitly driven by geopolitical concerns (cf. Rasizade 2002).

In this context, the period from 1999 to 2002 was a period of transition. At the 1999 summit of the Organisation for Security and Cooperation in Europe (OSCE) held in Istanbul, an agreement was signed between the presidents of Azerbaijan, Turkey and Georgia – what was called the Intergovernmental Agreement (1999) – to build what came to be known in the West as the BTC pipeline – or to many Georgians 'Bakuceyhan'. A year later, so-called Host Government Agreements were signed between the companies making up the BTC consortium and the three governments, thereby establishing a distinct legal regime for the route of the pipeline from Baku to Ceyhan. Following these two series of agreements, a new phase in the political history of the pipeline began. The transnational politics of the pipeline came increasingly to revolve not around the competing actions and interests of sovereign states, but around the specificities of pipeline routing, engineering, and social and environmental impacts. The International Finance Corporation (IFC) of the World Bank and the European Bank for Reconstruction and Development (EBRD), in particular, became critical to the governance of the pipeline, monitoring the project according to an emerging array of international principles. Indeed, in this period it became apparent that the pipeline was seen both by the IFIs and BP as a demonstration of how such principles might be put into practice. Indeed, the involvement of the EBRD and the IFC was doubly beneficial. On the one hand, the scrutiny exercised by the IFIs potentially reduced the environmental and reputational risks of the project for BP, as well as other major investors (Carroll 2012, cf. Mansley 2003). On the other hand, the sovereign states and state oil companies involved in the project, including Azerbaijan, required finance, but 'that could only come from the IFIs' (Browne 2010: 170). In these circumstances, the ordered practices of national environmental regulation, global governance, corporate responsibility and economic calculation appeared progressively to displace the chaotic and antagonistic realities of international politics (Walker 1993, 2010: 88).[8] If in the earlier phase international politics had been purified of any account of commercial interest, in this second phase the involvement of interested states largely disappeared from the transnational public debate.

Nonetheless, even in this second phase, after the agreements had been signed, national governments continued to participate in the politics of the pipeline. Through the involvement of the IFIs and the support provided through their own national export credit guarantee agencies, Western governments would come to have a substantial financial and political interest in the project. The US and UK governments, in particular, continued to monitor the project's progress via their own representatives on the IFC and EBRD boards, thereby further reducing the level of political and security risks for private investors (Watts 2005). In effect, the interests and powers of Western governments, which had been performed in public at the Istanbul Summit in 1999, now went backstage, to be held in reserve, ready to be deployed as and when necessary. Although the US and UK governments no longer played a public role in the project, in practice British government and embassy officials maintained a close interest in the development of BTC, while the US State Department established the post of ambassador specifically for the Caucasus pipeline system.[9] Whereas analysts had previously assumed that states had definite interests and concerns, once the decision to construct the pipeline was taken, the nature of the continuing involvement of Western states in the development of the pipeline became unclear. As we will see, the apparent movement from a regime dominated by the interests of states to one governed by national environmental law and the transnational norms of global governance could easily be reversed. But Western governments were not the only ones to remain politically involved, for the Georgian state was also, at specific moments, explicitly engaged in the pipeline's construction.

Physical Geography

In his prospectus on the 'scope and methods of geography', Halford Mackinder had argued that the development of a rational field of geographical inquiry should be grounded in knowledge of physical geography (Mackinder 1887: 214). But in the period prior to 2002, knowledge of physical geography did not appear to matter at all to those analysts interested in the question of how to export oil from the Caspian to the West. Given their focus on the actions and strategic interests of states, any understanding of, for example, the geology, climate or landscape of the South Caucasus in general, and of Georgia in particular, was of incidental significance to political analysts of the period. The technical and the material were taken to be external to the domain of politics.

From 2002, however, knowledge of the geography of Georgia began to play an increasingly visible part in the political life of the pipeline. By this point, four possible routes had been proposed through south and west Georgia. One 'western corridor' proceeded along the central valley of Georgia before turning south to cross the Turkish border. However, according to BTC's environmental impact assessment, 'this proposed corridor passes through extremely difficult rugged terrain along the upper reaches of the Mtkvari River Valley, where it also passes through the Borjomi-Kharaguali National Park' (BTC/ESIA 2002j: 4-9). The second, 'eastern corridor' took a more southerly route that passed further inland near the town of Akhalkalaki. Although this was considered to be an environmentally sound option by specialists (NCEIA 2003), it cut across an area where there was a large Armenian population and in which there was also, at this time, a Russian military base.[10] In these circumstances, 'the military facilities within the district of Akhalkalaki were considered to present unacceptable HSE [Health, Safety, and Environment] and other risks for pipeline routing and the Eastern corridor was discounted' (BTC/ESIA 2002j: 4-10, Jeter 2005: 5, Browne 2010: 167).[11]

With the 'eastern corridor' ruled out on the grounds of its insecurity, public debate focused on the feasibility and environmental impact of one of two central routes. One of these, the 'central corridor', had originally been selected by BTC; the other, a 'modified central corridor', took a route that was also 'based on the need to relocate the corridor' away from the vicinity of the Russian base. However, the modified central corridor passed near the winter resort of Bakuriani and the boundaries of the Borjomi-Kharagauli National Park: '79000 hectares of pristine forestlands and subalpine meadows ... formed the basis of the State Nature Reserve in Borjomi district, which has been under strict protection for decades' (Borjomi-Kharagauli National Park 2002: 5). The park, which is now a certified protected area, was considered to be 'one of the most significant parts of Georgia in terms of environmental, economic, cultural, and aesthetic considerations' (BTC/IEC 2006: 20). Equally significantly, Borjomi is a famous spa town developed by Tsar Nicholas II in the late nineteenth century, and home to a brand of mineral water widely known across the former Soviet Union, and a major source of export earnings for the country. At this time the head of the group that owned the Borjomi mineral water plant reputedly commented that, although he thought that nobody was opposed to the pipeline, 'I do not believe that there is no alternative route'.[12]

There appears to have been broad agreement amongst geoscientists that the modified central corridor was not ideal, due to the risk of landslides (BTC/ESIA 2002j: 4-26, Jeter 2005), and the question of whether the route posed a risk to the national park and to the purity and reputation of Borjomi mineral water came to be widely discussed in Tbilisi. Some opposition MPs criticised the Borjomi route, including a member of Mikheil Saakashvili's opposition National Movement,[13] who was reported to have said that 'all Georgians should stand together and defend Borjomi'.[14] At this time there was an extended live debate, with participants drawn from across Georgian politics, on the pro-opposition television station Rustavi 2. The World Wildlife Fund and the Georgian Ministry of Environment, which had commissioned a report from the Netherlands Commission for Environmental Impact Assessment, were also highly critical of the choice.[15] The Dutch Commission argued that there were three particular locations where

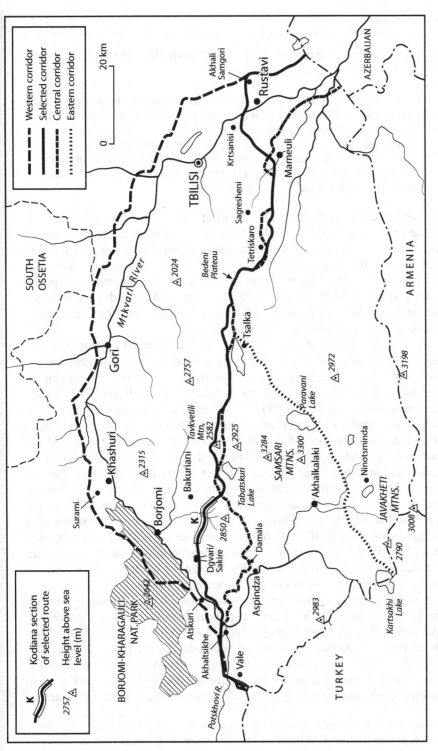

Figure 2.1 Alternative pipeline routes through Georgia. Map prepared by Ailsa Allen, School of Geography and the Environment. Reproduced by permission of the University of Oxford

landslides might cause damage to the pipeline and lead to oil spills, thereby posing significant risks to the environment:

> In determining the route of the pipeline] the pre-cautionary principle should be applied to the Borjomi/Bakuriani area because: (i) these resources do have considerable ecological as well as economic value; (ii) the ground water resources are protected; (iii) the surface water rivers, groundwater in the river valley alluvium, groundwater in Volcano-clastic formation are most vulnerable to oil spill pollution and (iii) [sic] for the ground water springs there is potential risk of oil spill pollution. (NCEIA 2002: 11)

This argument was reflected in the position taken by the Georgian Minister of the Environment, Nino Chkhobadze, who was reported to have argued that the risk of landslides in the Borjomi region had been insufficiently addressed by the BTC company in the ESIA report for Georgia. Moreover, she agreed with the Dutch Commission that a fifth, alternative pipeline route running through the Karakaia Mountain and Aspindza – which had not yet been investigated, and which involved the construction of a tunnel – was feasible and should therefore be considered.

Despite the Dutch Commission's and the Minister's criticisms, the Georgian government of Shevardnadze agreed to the 'modified central' route on 2 December 2002, despite Chkhobadze's reported misgivings.[16] According to the BTC company, the designated route, despite its difficulty, was both viable and secure (BTC/ESIA 2002i&j). Moreover, the project had been promoted by the Shevardnadze government as vital for Georgia, providing the basis both for energy security in the long term, and for employment and foreign investment in the short term. The Georgian Parliament speaker, Nino Burjanadze, was reported to have said that 'we cannot decline the offer to put the pipeline through Georgia ... as this is a real guarantee of our future security'.[17] Thus in December 2002, in the wake of this controversy, which was not fully resolved nor resolvable, the Borjomi route was finally selected (Green Alternative 2005). During what proved to be the last year of the Shevardnadze government, the relation between the pipeline and the physical geography of the country came to have a highly visible role in Georgian domestic politics.[18]

Nonetheless, the terms of the environmental permit granted to BTC by the Georgian government in 2002 stipulated a series of further conditions and 'continuing activities' (Georgia 2002, 2004: 83, BTC/ESIA 2004). They encompassed 'additional studies of landslide hazard areas' (BTC/ESIA 2004: A 1–2), more extensive consideration of biodiversity production (ibid.: A 1–3), and a series of activities that related specifically to what the Georgian government termed the 'Borjomi zone'. These included, most significantly:

> additional design and operational measures to secure the integrity of the pipeline in the event of third party intervention in a manner which allows sufficient time for information to reach project operations staff and State

security services to enable access to the Borjomi Area. Based upon risks which are foreseeable under the prevailing conditions, including risks associated with attempts to tap the pipeline, acts of vandalism, and attempts to disrupt the project by small organised groups, BTC co shall institute a programme [of activities]. (ibid.: A 1–3)

It is this clause in the government permit that subsequently came to be at the heart of a confrontation between the oil company and the Saakashvili government in July 2004: a crisis that erupted just eight months after the Rose Revolution that brought Saakashvili to power in Tbilisi. This crisis took the form of the government instructing the BTC company to stop pipeline construction as work proceeded in the Borjomi area in the summer of 2004, citing the terms of the environmental permit (Green Alternative 2005). In Tbilisi, speculation flourished about why this dispute had occurred at this time. Some informants viewed it in purely political terms, as part of a strategy by the new Saakashvili government to gain Western support in relation to challenges such as the growing tension over the region of South Ossetia. Others saw it, in contrast, as a symbolic gesture on the part of Saakashvili to defend the integrity of Georgia, thereby performing the sovereignty of the state in the face of foreign (Western) intervention (cf. Jeffrey 2013).[19] One informant saw the confrontation as an important act in which the Georgian government was seen to make good the 'betrayal' of Borjomi that had occurred in 2002.[20] But according to other informants, the late intervention by the Georgian government over the question of Borjomi reflected their limited capacity to carry out environmental assessment during this period (NCEIA 2004: 8).[21] As we shall see, the government's intervention occurred during a period in which villagers in the Borjomi region had also taken direct action to stop construction work (see Chapters 6 and 8).

Both BTC company and government insiders insisted, however, that this dispute – which led to the instruction to stop construction – was actually about matters of substance: how to address the existence of geohazards and security risks in the Borjomi region. It was, they agreed, about the technicalities of environmental protection in a sensitive region, and not about politics at all. At the heart of the clash over the government's environmental permit was a key point of material disagreement between the two parties. It revolved around the depth at which the pipeline should be buried. It had been originally agreed that the pipeline would be buried to a depth of one metre (Browne 2010: 168). However, the new Georgian government insisted that, in the Borjomi region, the pipeline should be buried more deeply in order to reduce the security risk identified in the environmental permit. The dispute led to the intervention of the US government, including a visit to Borjomi by US Assistant Secretary of State Elizabeth Jones. A week after the start of the Georgian government's actions, the matter was raised in meetings between Saakashvili and US Secretary of State Colin

Powell and even, according to newspaper reports, Defence Secretary Donald Rumsfeld.[22] In the ensuing negotiations a compromise of two metres was reached, leading Jones to declare that the question had become, once again, a purely technical matter: 'I am convinced', she was reported to have said, 'of the environmental integrity and sanctity of the pipeline. The security issues involved with the pipeline are being taken care of'.[23] In this instance, the US government saw itself as playing a part in restoring the integrity of an object that had been threatened, although not destroyed, by the intervention of the Georgian government. If the conflict between BP and the Georgian government appeared formally resolved, the episode had further consequences. At the end of summer 2004, a total of $6 million funding was given by BP for security measures, and a further $40 million for development projects, throughout Georgia (BTC/Georgia 2004).[24] Notably, the payment of the $40 million grant was conditional on the construction of the pipeline.

Despite the formal resolution of the 2004 dispute, the significance of landslides and soil erosion and the danger of sabotage in the 'Borjomi zone' had clearly been highlighted. But after the events of December 2002 and July 2004, the public debate about the Borjomi question declined. The controversy over the risks of taking the pipeline across the Lesser Caucasus became translated into new, more technical and complex forms, to be addressed by further geoscientific studies, by geotechnical engineering, by longer term monitoring and by minor adjustments to the route. From 2003, engineers and geoscientists had already begun to discuss the need for 'geotechnical engineering works at each of seven landslide sites' in the region (Chkheidze et al. 2005: 436). And at a conference held in 2004 at the Institution of Civil Engineers in London one month before Saakashvili's remarkable intervention, British and Georgian geoscientists working for BTC documented the importance of tectonics, seismicity, weather, rock and mudslides along the route through the Borjomi zone, particularly along the Kodiana section and in the Sakire area.

The BTC scientists were open about the specific challenges that the area presented due to the conjunction of a series of elements including cold conditions, fractured rocks, weathered silt-clay, intensive snow melt and northern slopes – which together conspired to produce a 'classical landslide environment' in the area, made worse by human activity 'including plough-ing, irrigation and forest clearance' (ibid.: 431). Landslide prediction was, as one of the conference contributors later noted, inevitably an inexact science, particularly given the lack of available data on the history of previous landslide events (Lee and Jones 2004, Lee 2009). Given the complexity of landslide phenomena, scientific prediction depended on the exercise of qualitative professional judgement, not on excessive reliance on precise metrics. Moreover, even to the extent that the likely location of future landslide activity could be judged, 'the management of landslide stability hazards along pipeline routes in mountainous areas [such as this] is very challenging. It requires input from experienced specialists across a range of disciplines including geomorphology, geology, geotechnical engineering, pipeline engineering, drainage and restoration' (Chkheidze et al. 2005: 444). Such management would entail the construction of, amongst other things, a series of containment facilities 'installed in riverbeds near to the Tskhratskaro-Kodiana section of the pipelines, as well as regular monitoring and incident reporting', as required under the terms of the Georgian environmental permit (BTC/PMDI 2006: 12, Georgia 2002, Marriott and Minio-Paluello 2012: 177). However, the efficacy of this technical solution could, in principle, be challenged, whether through further geoscientific research or, indeed, as a consequence of the unpredictable movement of land in the Kodiana section or the area around Sakire. If the resolution of the controversies of 2002 and 2004 had led to the folding of the political into the technical, it is also apparent that the physical geography of the Borjomi region remained of political consequence.

The Pipeline Regime

The controversy over the route of the pipeline through the 'Borjomi zone' revolved around the possible impact of pipeline construction across the region, and indeed the entire country, well beyond the immediate environment of the pipeline. The risk of landslides in Kodiana and Sakire and Saakashvili's dramatic intervention both threatened to destabilise the project as a whole. However, in principle, BTC expected that the impacts of pipeline construction could be contained within a restricted space running along the length of the pipeline, which was regulated not primarily by the Georgian state but according to a series of international guidelines and principles, and which was monitored by a series of different institutions and experts including consultants working for BTC and its lenders and specialists from the IFIs (BTC/ESAP 2002a). It was within this restricted space that BTC sought to manage, monitor and respond to the unpredictable movement of soil, pipes and trucks and the potentially unruly conduct of affected populations.

Indeed, a series of Host Government Agreements (HGAs), concluded in 2000 between Azerbaijan, Georgia and Turkey and the oil companies, indicated one of the key issues at stake in the development of the BTC pipeline (GYLA 2003). According to these documents, the pipeline would be associated with a distinct legal regime governing a narrow corridor of land running from Baku, across Georgia and Eastern Turkey, to Ceyhan. This would be a continuous territory that would not be contaminated by the inconsistency and variability of national legislation, the unpredictability of local legal proceedings, the risks posed by corruption, or the actions of regional authorities and national governments. According to these Agreements, 'in no event shall the Project be subject to any such standards [enacted by regional or inter-governmental authorities] to the extent that they are different from or more stringent than the standards and practices *generally prevailing in the international Petroleum pipeline industry for comparable projects*' (Host Government Agreement 2000b, article 4.4, emphasis added). In this way, the body of law relating to the route of the pipeline carved out a zone of regulation that could not be 'more stringent' than the standards and practices prevailing elsewhere, and that would be similar to the laws governing the construction and operation of other pipelines around the world: an index of the mobility and transposability of neo-liberal rule (cf. Barry 2001, Ong and Collier 2005, Ong 2007). Hannah Appel has argued, based on fieldwork in Equatorial Guinea, that the oil industry takes a modular form, replicating its working practices across a range of sites worldwide (Appel 2011, cf. Ferguson 2005). This observation applies well to the construction of the BTC pipeline, for example, in the organisation and population of employment. By the early 2000s, camps had been established along the route of the pipeline to house migrant workers from India, Latin America and North Africa who had experience of other infrastructure projects, and who were segregated from local populations. Local people, in turn, were employed largely as unskilled or semi-skilled workers, or in some instances as community liaison officers. The HGAs can therefore be described as modular agreements, establishing the legal basis for a separation between the route of the pipeline and the society and territory through which it passed, and thereby enabling the company to replicate the 'standards and practices' governing the conduct of the pipeline industry elsewhere (cf. Tsing 2009). In this way the distinction between the 'economy' (of pipeline construction) and the 'state' was enacted in a specific spatial and legal form (cf. Mitchell 1999).

The extent to which the pipeline corridor should be governed by its own distinct legal regime was, however, contested. Focusing their attention on the Host Government Agreement between the project participants and Turkey, lawyers working for Amnesty International in London asserted that the pipeline corridor would not be subject to the same human rights laws that might be adopted by Turkey in the future, as these progressively improved:

On the surface, the project is undertaken by a consortium led by BP that, in its planning, has taken steps to follow basic international standards for the protection of persons, property and the environment. Beneath the surface, the project in its day-to-day operation is excluded from certain important regulations by the state, even when these would translate international standards into Turkish law ... [As a result] while Turkey remains bound by its international human rights obligations, it has undertaken in the Host Government

Agreement (HGA) to pay the consortium substantial compensation for any changes in law or other actions that will disturb the equilibrium of the project. (Amnesty International 2003: 5, see Host Government Agreement 2000c)

Responding to Amnesty's criticism of the HGA with Turkey, the BTC company insisted they had no intention of preventing Turkey from honouring its international human rights obligations as they continued to evolve. Nonetheless, the company issued a further statement, offering the clarification that the HGAs and 'other project agreements ... provide a dynamic benchmark that will evolve as EU standards evolve' (cited in Redgewell 2012: 109, Browne 2010: 171). The legal regime governing the operations of the pipeline corridor would therefore be compatible with some elements of EU law, but it nonetheless remained distinct, enacting the BTC company's intention to forge a separation between the corridors within which the pipeline would be constructed and the territory of the state through which these corridors passed (Redgewell 2012: 106-107). Critics argued that, despite the clarification, a series of legal questions remained unresolved (Baku-Ceyhan Campaign 2003a: 3).

At the same time, throughout 2001–3, the route of the BTC pipeline became an increasingly intensive zone of information production. Initially this involved a range of experts with backgrounds in fields including pipeline engineering, geology, ecology, geo-hydrology, social impact assessment and public participation. Specialists from the international consultancy firm Environmental Resources Management (ERM), for example, working with consultancy firms and universities in the region, carried out an Environmental and Social Impact Assessment (ESIA). It is this period of intensive information production that led in 2002 to BP making public the initial documents of what by mid-2003 had become a vast archive of documents running to 38 volumes and thousands of pages,[25] informed by the principle of transparency and responding to the demands of the IFC and EBRD who were intending to provide financial support for the project. The organisation of this archive, which I discuss in more detail in later chapters, embodied the complex division of labour between a growing range of technical experts.

While the pipeline's legal regime was contested by Amnesty International's lawyers in London, the archive was interrogated further by other metropolitan intellectuals and researchers based in London, Prague, Tbilisi and elsewhere. In Tbilisi, the Georgian NGO Green Alternative, supported financially by Oxfam and forming part of the Central European Bankwatch network, had begun an increasingly visible campaign, insisting that the oil company should follow international guidelines to the letter while encouraging and assisting individual villagers to take their complaints directly to the international financial institutions (Green Alternative 2002, Green Alternative 2004a&b, Green Alternative et al. 2004, Green Alternative/CEE

Bankwatch 2005). While there is a long tradition of environmental politics in Georgia, which developed within the Soviet system, Green Alternative directed their politics less towards the Georgian state than towards the practice of ethical or environmentally responsible capitalism. In Turkey and Azerbaijan, in contrast, public debate about the construction of the pipeline was much more limited, reflecting the limited development of environmental politics, as well as the strength of the state and its capacity to control dissent in both countries (O'Lear and Gray 2006, O'Lear 2007, Cheterian 2010, see also Oldfield et al. 2004). In Azerbaijan, it reflected also the comparative lack of US and Western European financial support for local NGOs (Hamilton 2004). Nonetheless, a handful of activists gave publicity to both Turkish and Azerbaijani villagers' complaints about the operation of compensation payments, amongst other matters (e.g. OWRP 2004).

Meanwhile, in Western Europe, a diverse and increasingly vocal group of activists, researchers, lawyers and artists associated with an array of NGOs and protest groups, including members of Friends of the Earth International, the Kurdish Human Rights Project (KHRP), La Campagna per la Riforma della Banca Mondiale, The Corner House and Platform, formed the Baku-Ceyhan Campaign (Platform et al. 2003: 4). The campaign began to conduct a series of short 'Fact-Finding Missions' to Georgia, Turkey and Azerbaijan, drawing on the support of Georgian NGOs, including Green Alternative and the Georgian Young Lawyers Association (GYLA), as well as NGOs in Azerbaijan and north-east Turkey (Green Alternative et al. 2002, Bank Information Center et al. 2003, Campagna et al. 2003, CEE Bankwatch et al. 2004, Centre for Civic Initiatives 2004, 2005a&b). In the context of a public consultation process initiated by the IFIs in mid-2003, the Baku-Ceyhan Campaign argued that there were systematic gaps between the international agreements and guidelines governing the development of the project and their practical implementation. Drawing on their accounts of the findings of their missions and a detailed examination of BTC project documents the campaign contended that these gaps were both undeniable and wide-ranging: indeed the project 'continues to breach all relevant World Bank safeguard policies on multiple counts, in addition to violating other project standards' (Baku-Ceyhan Campaign 2003a: 2). These standards related to an extraordinarily wide range of issues including *inter alia* environmental assessment, natural habitats, consultation, indigenous peoples, involuntary resettlement and cultural property. The campaign observed, moreover, that those commercial lenders that had adopted the Equator Principles were committed not to provide loans to projects that did not meet relevant World Bank and IFC social and environmental standards.[26] In these circumstances, the project clearly breached 'the letter and the spirit of the Equator Principles' (Baku-Ceyhan Campaign 2003c: 3). There was also evidence, they contended, of breaches of European Union Directives, the force of which was recognised by the

Istanbul agreement, EBRD environmental policy and the OECD guidelines for Multinational Enterprises. Given the extent of these violations, the campaign argued that the IFIs and commercial lenders should insist that the project conform to the principles and standards to which they themselves had committed. In short, not only did the BTC company need to demonstrate its ethical credentials, but so too did the IFIs and the project's commercial lenders (Baku-Ceyhan Campaign 2003c: 4, Chapter 4).

The Director of Friends of the Earth put the Baku-Ceyhan Campaign's criticisms of the project in more general and stark terms: 'DFID should not be financing British companies like BP who are profiting from dirty energy schemes that contribute to climate change and to environmental and social destruction'.[27] Anticipating an emerging conflict between the BTC company, the IFIs and the NGOs over the conduct and contents of ERM's environmental and social impact assessment, the eco-protest group Rising Tide occupied the offices of ERM claiming that the consultancy firm were playing a 'crucial if low-key role' in 'grooming BP's Baku-Ceyhan pipeline for public investment': 'We're in their office today to expose that climate change-inducing role, one which is defined more by the profit-motive than any desire to improve the environment or make life better for communities affected by the oil industry' (Rising Tide 2002).[28]

In effect, the response of the Baku-Ceyhan Campaign to the apparently overwhelming quantity of detailed information made public by the oil company was to provide in return an overwhelming response. Their disputation of the vast numbers of claims made in the published archive took in not just the legal provisions of the HGA and the conduct of the ESIA and public consultation practices, but the resettlement of populations, corruption and biodiversity, as well as the broader impact of escalating oil production on climate change (Platform et al. 2003, Marriott and Muttit 2006, Hildyard and Muttit 2006). In relation to the IFC operational policy on natural habitats, for example, the campaign argued that there was 'no evidence that local communities have or will play significant roles in planning, designing, implementing or monitoring project in relation to natural habitats' (Baku-Ceyhan Campaign 2003a: 18). In short, the project simply did not comply with the relevant IFC Operational Directive. During 2002–3, the war of claim and counter-claim progressively escalated between the Baku-Ceyhan Campaign NGOs, the UK government, the IFIs and other lenders over whether the development of the pipeline had breached international guidelines and principles.[29] Given the alleged range and seriousness of the BTC company's breaches of such guidelines and principles, the campaign NGOs contended that neither the IFC nor the EBRD nor the UK government should support the project. During this period, officials in the UK Department for International Development joked that the Department had to deal with more correspondence about BTC than about the Iraq war.[30] The comparison with the Iraq War was in certain respects apposite, for in 2003 in London the politics of both the Iraq War and the BTC pipeline had come to centre on disputes involving matters of fact and the relation between these disputes and interpretations of international law. The arguments made by the Baku-Ceyhan Campaign not only received coverage in the media but gained some influence in mainstream political circles. Thus, echoing the Baku-Ceyhan Campaign's position, the Minister for the Environment in the Labour government in 2003, Michael Meacher, commented later that 'a huge new oil pipeline, opened a week ago but not fully operational till August, is set to become an environmental, political and economic time bomb. Over 1,000 miles long, it is a classic example of pretensions to corporate social responsibility claimed by the BP consortium being trampled all over by the stampede for oil' (Meacher 2005).[31]

In late 2003, however, having considered representations critical of BTC's alleged failures to meet relevant international standards and guidelines made by a range of NGOs, officials from the IFC and the EBRD dismissed their claims (IFC 2003a). Shortly thereafter the IFIs' boards decided to give financial backing to the construction of the pipeline (IFC 2003b). According to the IFIs, the facts presented in the NGO reports were simply not facts, and their claims concerning the violation of the World Bank and EBRD guidelines were unsubstantiated.[32] Claims made by the campaign about specific villages did not demonstrate that the company had failed to fulfil the terms of the World Bank operating principles. In short, the IFIs took the NGOs' facts to be, as November et al. put it, 'spurious referents' which possessed 'no practical counterpart' (November et al. 2010: 117).

This conclusion, in turn, led to a stinging response from members of the Baku-Ceyhan Campaign based on conclusions they had drawn from their own fieldwork. The IFC, it was claimed, 'consistently fails to answer allegations directly; where NGOs have provided evidence that a particular policy is not being implemented properly, it merely redirects attention back to the paper policy without any engagement with the reality on the ground. It is largely toothless, relying on hortatory banalities (such as the "BTC Co clearly recognises the importance of being diligent")' (Baku-Ceyhan Campaign 2003d: 1–2). Increasingly, it was not just the conduct of the ESIA that was disputed by the campaign's coalition of international NGOs, but the claims to knowledge made by BP and the IFIs about 'reality on the ground'. Yet despite the decision made in December 2003 by the IFC, the EBRD and the UK government to provide financial support for the project, the numerous, proliferating controversies were not resolved. The campaign continued to conduct Fact-Finding Missions along the length of the pipeline until September 2005,[33] generating what was claimed to be further evidence that refuted the accounts given by the BTC company and the IFIs. In short, the accumulation of evidence by both sides of the disputes led not to a resolution of their differences, but to a sense of the impossibility of their resolution. What might have appeared to be a series of agonistic disputes over matters of fact and the interpretation of international guidelines and agreements was increasingly conducted in antagonistic terms (cf. Mouffe 2005a). The commitment to transparency on the part of the international organisations and BP was mirrored by the campaign NGOs' commitment to the production and publicisation of research-based counter-claims to matters of fact (cf. Eden et al. 2006).

Conclusion: Political Knowledge

In criticising Trotsky's justification for the Soviet Invasion of Georgia, Bertrand Russell had forcefully argued that the Russian intervention was all about oil, and not about socialism at all. In recent years, however, the philosopher's reductive explanation of the value of Georgia for foreign powers has been rediscovered. As I have shown, since the early 1990s, with the collapse of state socialism, Georgia came to be seen as having strategic importance for the West because it lay on the route from the oil fields of the Caspian, the control of which was thought vital for Western energy security. During this period, the BTC pipeline was represented primarily as a line on a map, its route explicable in terms of the geopolitical context through which it passed.

Russell's participation in the Trade Union delegation to Russia in 1919, however, points to the importance of a mode of political research and analysis that is not focused on the strategic interests of states. In seeking to examine conditions in Russia, Russell was more concerned with the question of the relation between socialist theory and the reality of bolshevism than with matters of international relations and foreign policy. Whatever the evident weaknesses of his writings, he sought to interrogate the practice of bolshevism through direct observation and discussion – in other words, through political fieldwork – as well as through the inspection of canonical socialist texts. Clear parallels can be drawn between Russell's research in Soviet Russia and the work of more recent visitors to the Caucasus. For, from the early 2000s on, the 1760 km route of the BTC pipeline across three states became the site for a range of styles of fieldwork carried out not just by oil company consultants and specialists working for IFIs, but also by delegations from the Baku-Ceyhan Campaign as well as other individuals and NGOs. All of these researchers were concerned, for different reasons, with the question of the relation between the principles of transparency and corporate responsibility, as they were embodied in a series of international agreements and guidelines, and their enactment in the field. They were interested, in other words, in the relationship between the theory and the practice of an emerging form of capitalism – one in which business is expected to perform ethically (Barry 2004, Watts 2005, Thompson 2012, Chapter 4). While Russell turned his attention to interrogating 'the failure of Russian Industry' and the real meaning of the dictatorship of the proletariat, the political fieldworkers of the early 2000s sought to assess the successes and failures of the oil industry in enacting the principles of transnational governance. In this way, the pipeline was political – as Deleuze put it – 'before being technical' (Deleuze 1988: 39). The generation of political knowledge was inscribed from the very outset in the construction of the pipeline.

In describing a potential future for the discipline of geography, Mackinder asserted that 'the function of political geography is to detect and demonstrate the relations subsisting between man and society and so much of his environment as varies locally ... [and that] it is the function of physical geography to analyse one of these factors, the varying environment' (Mackinder 1962: 216). But this chapter has advanced a core reason for radically reworking Mackinder's proposal: knowledge of physical geography – and the geosciences more broadly – cannot be regarded as an uncontested foundation on which analyses of politics or society can subsequently be based. The knowledge claims of the geosciences can be uncertain; they can also become matters of public dispute, as in the controversy over the Borjomi route. In this sense they may form part of what I have termed a political situation, one that the political geographer seeks to analyse. Indeed, as I have shown, the question of the relation between the pipeline and its environment was critical to the politicisation of the pipeline route, in the region of Borjomi and elsewhere. Subsequent chapters, reflecting on a further series of disputes along the length of the Georgian section, suggest a second reason for reworking Mackinder's proposal. For as we will see, any knowledge of the environment that 'varies locally' must include not just an account of the natural environment, but one that also addresses the properties and behaviour of artefacts such as pipes, pumping stations, communication systems, construction traffic and access roads.

In the next chapter I turn from a consideration of the Georgian route to an account of the significance of the principle of transparency. I focus not on the BTC pipeline, but on the implementation in Azerbaijan of the international Extractive Industries Transparency Initiative (EITI). EITI throws light on how the question of transparency (or lack of transparency) has come to be critical to the governance and politics of oil. At the same time, addressing the modest scope of EITI in Azerbaijan throws into relief the far more ambitious and extensive experiment in transparency that developed around the BTC pipeline.

Chapter Three
Transparency's Witness

In his essay 'The Secret and the Secret Society', Georg Simmel noted the progressive openness of the state or, as he expressed it, 'publicity's invasion of the affairs of state'. This invasion had occurred to 'such an extent that, by now, governments officially publish facts without whose secrecy, prior to the nineteenth century, no regime seemed possible' (Simmel 1950: 336). Today, the question of the openness of the state has been framed in a particular way; for many the idea and practice of *transparency* has become critical for efforts to promote good governance. Transparency is a term, according to Christopher Hood, that has attained quasi-religious significance in debate over governance and institutional design: 'Since the 1980s the word has appeared in the litanies of countless institutional-reform documents and mission statements ... it is the pervasive jargon of business governance as well as that of governments and international bodies, and has been used almost to saturation point in all of those domains over the past decade' (Hood 2006: 3). Hood traces the demand for openness in government back to the work of Spinoza, Rousseau and Bentham. Bentham, in particular, drew an opposition between publicity and secrecy for 'the best project prepared in darkness, would excite more alarm than the worst, under the auspices of publicity' (Bentham 1839: 310).

If the recent enthusiasm for transparency is simply the latest stage in a long evolutionary process, of 'publicity's invasion', then what is new? Certainly, there is the prevalence of the term's usage. Hood does not seek to explain why transparency (rather than, say, openness) has become a

Material Politics: Disputes Along the Pipeline, First Edition. Andrew Barry.
© 2013 John Wiley & Sons, Ltd. Published 2013 by John Wiley & Sons, Ltd.

preferred term, but he does make an important observation: although the term has become pervasive it has also been promoted as a critical element in the recent development of transnational economic governance. In this respect, the oil economy is not unique in being subject to the claim that transparency might help address some of the difficulties and criticisms that it faces and the risks that it poses (Fisher 2010, Garsten and de Montoya 2008, Gupta 2008, Hajer 2009, Hood and Heald 2006, Jasanoff 2006b, Neyland 2007). The global finance industry has equally been a focus for calls for greater transparency (Best 2005, 2007), while national governments and international bodies such as the International Monetary Fund (IMF) and the World Trade Organization (WTO) have also sought to demonstrate that they are more transparent than before (Woods 2001, Larner 2009). But what is unique about the case of oil, I would suggest, is the stark contrast between the opacity of the way in which the multinational oil industry is said to have operated in the past, and the transparency of the manner that it is expected to act in the future.

One feature of transparency is that it is applicable not just to the activities of governments but also to the operations of all organisations, including business. Moreover, according to its supporters, the progressive extension of the application of the principle should not be a threat to commerce, nor does it necessarily entail a restriction on commercial activity, or 'the invasion' of publicity into the world of business. On the contrary, the implementation of transparency is said to provide the basis on which the information necessary for the proper function of free markets would become readily available. In this way, the practice of transparency has acquired a series of functions and multiple meanings (Grossman et al. 2008). First, transparency is expected to allow investors to make rational judgements concerning the strength of both commercial and public organisations, without having to gain access to insider knowledge. In these circumstances, transparency needs to be actively promoted. Indeed, 'a generation of economists has shown that markets and deliberative processes do not automatically produce all the information people need to make informed choices concerning goods and services' (Fung et al. 2007: 6). Secondly, the operation of the principle of transparency appears to establish a distinction between a domain within which more or less free markets exist and a domain external to their operation (Callon 1998b). In effect, the enactment of transparency is expected to establish a boundary between the legitimate domain of commerce and the market on the one hand, and the illegitimate territory of corruption and state crime on the other. Transparency is necessary if corruption is to be reduced, information is to flow more freely, organisations are to be held accountable, and free markets are to flourish.

Thirdly, in so far as it is directed towards the activities of governments, transparency is thought to foster public accountability for 'effective accountability requires mechanisms for steady and reliable information and communication between decision-makers and stakeholders' (Held and Koenig-Archibugi 2005: 3). It is not just a matter of making information public, but a matter of moulding institutions into forms that are able to perform it (Power 1997, 2007a, Strathern 2000). Transparency is thus expected to foster the development of the kind of persons and institutions that are in a position to use and assess the credibility of any information that is published. The operation of transparency, thus, should lead not just to the production of information, but transformation in the identities and conduct of persons and organisations. It is, in short, a technique of governmentality, a device intended to 'articulate actions: [to] act or [to] make others act' (Muniesa et al. 2007: 2).

Simmel's essay, however, points to two further key issues. One is indicated by his metaphor of invasion to describe the effect of openness on the state. In describing openness in these terms, Simmel complicates the terms of the opposition between transparency and secrecy, and tradition and innovation. For Simmel, it was not openness, but secrecy that was 'mankind's greatest achievement'. Increasing openness and increasing secrecy *both* demanded innovation, in opposition to a past in which neither concept had so much salience. Openness did not reduce secrecy, he argued, but intensified the demand for it. 'Real secrecy', Simmel argued, only began historically with the development of greater openness. Yet, in his view, secrecy was valuable, while openness was not: 'In comparison with the childish stage in which every conception is expressed at once, and every undertaking is accessible to the eyes of all, the secret produces an immense enlargement of life: numerous contents of life cannot emerge in the presence of full publicity' (Simmel 1950: 330). The development of practices of openness does not, therefore, reduce a given reservoir of secrets. Rather, it transforms the nature of what is kept secret, and what is valuable to keep secret and what is not. Moreover, the development of practices of openness coincides with the development of practices of secrecy. Indeed, this is not surprising. For when so much is out in the open, what is not acquires a new and arguably greater value (Strathern 2000: 310). Conversely, the growing prevalence of secrecy heightens the importance of openness, channelling attention towards objects that were invisible but subsequently come into public view (Taussig 1999: 56).

Secondly, as Simmel observes, in a world of openness, it is not just openness but secrecy itself that has to be achieved. Secrecy may rely on the use of law, through technology (such as screens and firewalls), through economic control, through the deployment of cultural capital, or through the threat of violence. Just as transparency is associated with certain assemblages of persons and objects and devices (such as a free press, accounting procedures and disclosure requirements, public reports, debates and forums, etc.), promoted through the work of particular institutions, there also exist devices of secrecy. Simmel's essay points us towards an interest in the history of such devices and the interconnections between the histories of openness and secrecy. Like deviance for Foucault, corruption, bribery, illegality and so on are implicitly spoken about in practices of transparency, but often through their absence, or in the margins of texts (Foucault 1979).

It would be a mistake, then, to assume that greater transparency simply leads to less secrecy, or vice versa. In what follows I make two sets of pre-liminary observations. The first is that the practice of transparency raises questions and may lead to passionate disputes not only about what is and what is not published, but about the processes by which public information is generated. As sociologists of scientific knowledge have shown, published scientific papers provide a very limited report of the messy processes of the research that they purport to describe (Collins 1985, Law 2004). In effect, they direct the reader away from what has been called the circumstances of their production (Latour and Woolgar 1986: 240, Latour 1987). The same can be said, as we shall see, of the kinds of reports produced in response to the demand for transparency. The exercise of transparency may lead to questions about how the information that is published has been produced or circulated (see Strathern 1991: xxii). Critics may try to uncover the truth about the process that has generated what has been made transparent, acting in the manner of critical sociologists of science. But in turn, those who attempt to uncover the truth that lies behind the façade of transpar-ency may provoke questions about the lack of transparency of their own practices. In effect, the publication of information about matters that have previously not been made public generates further candidate secrets. As Deleuze and Guattari observe, those who are expert in finding secrets nec-essarily have to maintain some secrets themselves. In this way, 'from an anecdotal standpoint, the perception of the secret is the opposite of the secret, but from the standpoint of the concept it is part of it' (Deleuze and Guattari 1987: 287). Both those who promote transparency and those who interrogate the limits of transparency will inevitably leave a great deal that is undisclosed about their own practice.

A second observation is that transparency points inevitably to the existence of a domain of activity about which it is thought that information has not yet been or might never be made public, whether intentionally or not (Simmel 1950, Stoler 2009: 3, Gross 2010). Gaps may become apparent between that which is rendered transparent and that which may or may not be widely known but which, it is believed, will never officially become public knowl-edge. Thus, instead of having the effect of reducing the finite quantity of mat-ters that are not made public, the operations of transparency have the potential to highlight the existence of a vast range of matters that never will be made public (Barry 2006, McGoey 2007, 2009, MacIntosh and Quattrone 2010, Hetherington 2011), including matters that will not even be accessed by the most skilful social researchers (Quattrone 2006). Formal demands for transparency are therefore likely to co-exist with the circulation of rumours and 'public secrets' about matters that are not and may never officially become public knowledge, but which are nonetheless widely known (Taussig 1999). In the course of this research I was made privy to numerous such rumours and public secrets, some of which are probably true and many of which are probably not, but which cannot in either case be published. Whereas the information produced to meet the requirements of transparency is traceable, and therefore is expected to render institutions accountable, the origins of public secrets, as well as specific rumours, are indeterminate (Kwinter 2001: 126). The exercise of transparency does not reduce the importance of rumour, but rather gives it a new yet unacknowledged significance.

If the virtues of the principle of transparency are widely diffused, the principle is thought to have particular relevance to the extractive industries in general, and the oil industry in particular. Why has the principle of transparency acquired so much significance in relation to the politics and economy of oil? According to many commentators, countries with abundant oil resources have often failed to achieve good or sustainable levels of economic growth (Auty 1993, 2005, Sachs and Warner 2001, Le Billon 2005). Indeed, given the ease with which wealth can be extracted through the production of oil, the governments of oil-rich states have often been able to maintain public acquiescence by increasing state expenditure to unsustainable and spectacular levels in the short term (Coronil 1997, Karl 1997, Bannon and Collier 2003). The possession of this valuable resource has therefore not generally led to broad-based economic development and the establishment of democracy, but has all too frequently been correlated with economic stagnation, state repression, an absence of democratic freedoms and civil war: 'the link between natural resource abundance and the propensity for civil strife is now well established' (Auty 2005: 29; Collier and Hoeffler 2005, Humphreys et al. 2007). This state of affairs is known as the 'resource curse'. To be sure, the existence, prevalence and causes of the resource curse are all contested (Rosser 2006: 58, Ross 2001, Haber and Menaldo 2011). For some analysts, the presence of abundant natural resources such as oil leads to irrational behaviour, while for others it leads to what are claimed to be rational forms of economic behaviour such as corruption, looting, or the provision of financial support for *coups d'états*. According to some writers, an abundance of oil resources fosters foreign military intervention, while others stress how it may lead to rising political risks for foreign investors (Jensen and Johnston 2011) or is likely to weaken state institutions (Karl 1997).

Yet whatever the causes and form of the resource curse, most commentators agree that the promotion of greater transparency provides a good way of beginning to address it (Collier and Hoeffler 2005: 632, Humphreys et al. 2007, Swanson et al. 2003, Le Billon 2005: 24). For if the oil economy can be become more transparent, it is argued, then two critically important and desirable developments will follow: it should prove easier to reduce the level of corruption; and civil society will possess the information required to hold the governments of oil-producing countries to account. In this context, transparency appears to function primarily as a 'market device' (Callon et al. 2007). It is expected to make possible certain forms of economic calculation, while reducing the likelihood of those non-market forms of economic calculation associated with corruption and violence.

The arguments of the economists of the resource curse have been influential as we shall see in the first part of the chapter. However, in the second part of this chapter I argue that, in relation to the extractive industries, the project of transparency – the Extractive Industries Transparency Initiative (EITI) – has been understood to be as much a practical experiment in normative political theory as in economics. The aim of those who promote the virtues of transparency is not just to address the lack of economic information, and to foster the development of a market economy, but also to address the lack of development of civil society. The operation of revenue transparency entails not just the development of literary devices (displays of revenue payments, for example), but also the presence of witnesses, the cultivation of forms of ethical conduct through seminars, guidebooks and forums, as well as the existence of appropriate institutions. At the same time, in relation to both cases, transparency takes the form of a public experiment (Schaffer 2005), in which the witnesses to the experiment are not just economists and political theorists, but also 'civil society', 'stakeholders', auditors and the international community. In short, the development of transparency is expected to lead to the formation of a society and, in turn, this society will foster its progressive development in the future.

Moreover, particular public experiments in transparency are intended as exemplars which can be subsequently imitated elsewhere. A global society, concerned with the issue of revenue transparency, is formed through the replication and adaptation of a local model (Tarde 2001 [1890], Latour 2005b, Barry 2006). In effect, transparency operates along the borders between economic and political life. On the one hand, the implementation of transparency is expected to effect a form of politicisation of the economy that is measured, limited and rational. On the other hand, revenue transparency is intended to channel disagreements towards the specific question of economic calculation. A story that economists tell about natural resources – 'the resource curse' – is taken to have profound and wider implications for the public politics of knowledge.

Although EITI represents a systematic attempt to render the oil economy more transparent, it cannot be taken as representative of the operation of transparency in the oil industry in general. For while the EITI has been directed towards quite specific, limited and technical issues, the transparency of the BTC pipeline turned out to be much more wide-ranging and ambitious, involving a vast range of different experts and forms of knowledge production. Moreover, whereas the transnational experiment of revenue transparency occurred under carefully controlled conditions, both in Azerbaijan and elsewhere, the transnational experiment in transparency associated with the BTC entailed the formation of a more complex and evolving series of political spaces, stretching across three countries, and along a route 1760 km in length.

Resource Curse

In 2003 a team of economists from Columbia University visited the West African state of São Tomé e Príncipe (STP) in order to consider the impact of oil revenues on the economy.[1] This visit might seem surprising. After all, São Tomé, a small island state with approximately 160,000 inhabitants had yet, at this time, to receive any revenues from the production of oil. Indeed, following the exploration of offshore fields lying between STP and Nigeria, there was an expectation that oil production might begin as soon as 2012, or even 2010, but by 2007 this expectation had faded (Weszkalnys 2011).

Gisa Weszkalnys has argued that, despite having no revenues from oil production, STP might be regarded as an exemplary oil state. Together with Nigeria, STP signed the Abuja declaration (2004), committing both countries to transparency in relation to the countries' Joint Development Zone. On the advice of the Columbia University economists, the United Nations Development Programme (UNDP), the IMF and the World Bank, STP also adopted a Petroleum Management Law and a National Oil Account in order to invest oil revenues for future economic development. It had, in other words, anticipated the arrival of the oil economy and acted to prevent its potentially negative consequences. The transformation required was more than merely institutional. As Weszkalnys notes, 'anticipatory activities have not stopped on the level of the state, the law or institutional reform. What is especially needed ... is the creation of civil society and good governance, including a "change in mentality"' (Weszkalnys 2007: 3, see also Weszkalnys 2008).

While STP has yet to receive any oil revenues, the Columbia University project made it clear that it could, in certain respects, become a model to be replicated elsewhere. As well as the Petroleum Management Law, the project led 'to the design and execution of a National Forum, through which to inform citizens about the country's oil revenues and to solicit their views on how they might be spent'. The project also precipitated 'the formulation of a plan of action for sustainable economic development' and the publication of a substantial volume (co-authored by, amongst others, the Nobel Prize-winning economist Joseph Stiglitz, and Jeffrey Sachs, Director of the Columbia University Earth Institute) which offered a theoretical and practical guide to other countries with substantial natural resource wealth.

The book contained the Petroleum Management Law as a template for others to consider. Central to the claims made for the STP model was the idea of openness. The problem that STP could avoid was clear enough:

> The central problem facing resource-rich countries may be easily stated: various individuals wish to divert as much of that endowment as possible for their own private benefit. Modern economic theory has analyzed the generic problem of inducing agents (here government officials) to act in the interests of those they are supposed to serve (the principals, here citizens more generally). Agency problems arise whenever information is imperfect, and hence there is a need to emphasize transparency, or improving the openness and availability of information in the attempt to control corruption. (Humphreys et al. 2007: 26)

Although the idea of transparency is widespread, economic analyses of the so-called 'resource curse' (Auty 1993, Bannon and Collier 2003) have provided it with an influential justification. In this argument, countries possessing a wealth of non-renewable natural resources (typically associated with the oil, gas and mining industries) experience a series of problems that frequently lead to lower rates of growth than those occurring in countries with smaller endowments of natural resources. The Columbia University authors insisted, in particular, that the presence of such resources induced rent-seeking behaviour on the part of governments and elites. At best, this was likely to lead to a lack of investment and interest in other sectors of economic activity including, for example, agriculture and manufacturing, as well as public services such as health and education. It meant, furthermore, that states would be less reliant on taxation revenues and 'when citizens are untaxed they sometimes have less information about state activities and, in turn, may demand less from nation states' (Humphrey et al. 2007: 11). Moreover, rent-seeking opportunities frequently lead to widespread corruption and, in many cases, violent conflict, as different groups or foreign governments seek to gain control over revenues. Given these opportunities, the conduct of a *coup d'état*, for example, can be understood as a form of rational economic action. Indeed, one of the Columbia authors identified no less than seven distinct mechanisms leading to natural resource conflicts: the 'greedy outsiders mechanism', three variants of the 'greedy rebels mechanism', the 'grievance mechanism', the 'feasibility mechanism', and the 'weak state mechanism' (Humphreys 2005: 533).

The Columbia University project was not primarily an analysis of a specific example. Rather, it was a modest and practical intervention, based on the analysis of the resource curse, which formed an element of a much wider series of efforts, also involving the World Bank, intended to improve petroleum governance in STP. In the context of this analysis, the Columbia team had little to say about the colonial and post-colonial political history of STP: history remained outside the frame of economic analysis (Callon 1998b). Commenting on a *coup d'état* by army officers in July 2003, for example, Ricardo Soares de Oliveira notes that 'most analysts gave exclusive coverage to the perceived linkage with the oil contracts (a view aided by the coup spokesman's constant references to oil and social justice) and all but forgot the country's coup-prone past and the older grievances of São-Tomean society' (Soares de Oliveira 2007: 239, Weszkalnys 2009). Indeed, recent discussions of STP focus almost exclusively on oil, failing to attend to other aspects of political and economic life in the country, or the ways in which the notion of the resource curse itself intersects and draws upon 'familiar ideas about, and instances of, illicit wealth, appropriations of state property, or simply seemingly self-perpetuating patterns of social inequality' (Weszkalnys 2011; 366). In short, the strength of the Columbia approach did not derive from its attention to the specificity of STP and its history, but rather from its effort to transform the São Tomean economy into a particular example of a more general problem (the resource curse), and as a test site for a set of devices that were expected to reduce the level of resource conflicts not just in São Tome, but wherever they occurred. In effect, STP was conceived of as something of an 'island laboratory' (Greenhough 2006) for natural resource economics, an experimental site that would come to demonstrate the effectiveness of economic analysis in practice (Mitchell 2005: 297).

Disclosure

São Tomé is not an isolated case, however. It is one of 23 countries signed up to EITI, originally launched by British Prime Minister, Tony Blair, at the World Summit on sustainable development in Johannesburg in 2002. The initiative promoted transparency as central to the solution of the economic and political problems associated with the development of the oil, gas and mining industries in developing countries. Technically, EITI operates according to a principle of double disclosure: governments are expected to disclose what payments they receive from the extractive industries, and the extractive industries disclose what they pay to governments. These payments can be made through a variety of means. For example, EITI reports for Azerbaijan break down payments into the following categories: monetary inflow as government's entitlement in foreign companies' production stream; payments in kind (of both crude oil and gas) expressed in barrels of oil and cubic metres of gas; bonuses; transportation tariffs; acreage fees; royalties; profit taxes; other taxes, as well as taxes paid by local Azeri companies (Moore Stephens 2007). In practice, the payments made under each of these separate headings are aggregated. This means that nothing is recorded regarding payments to specific companies unless there is agreement from both the government and the companies concerned. Published figures do not reveal whether the Azerbaijan government sold oil and gas back to the companies prior to delivery, resulting in an increased cash revenue payment. Moreover, if this were the case, the price paid by the oil companies to the government is not recorded (EITI 2012a).

The demands of EITI are, therefore, very specific. They record one set of transactions in the circulation of natural resource revenues according to international accounting standards. They do not say anything either about the expenditure of such revenues by government or about NGOs funded directly or indirectly by the extractive industries. Nor do they record payments (in cash or in kind) along the oil production value chain such as payments to local subcontractors. Furthermore, while auditors provide an account of any discrepancy between the figures provided by governments and companies, this does not necessarily imply that all the figures are accurate or complete. If the aim is to reduce imperfections in the availability of information, then the reduction achieved is real but also quite modest. As a representation of the natural resource economy of a nation-state, the reports provided by the EITI process have been limited.[2] Given this observation, a number of countries might stand out as exemplary in their transparency, or in the level of their aspirations to greater transparency in the future. One is Mongolia, which was said by an informant to have produced a particularly clear way of presenting information. Another is Timor-Leste, which has embraced the idea of revenue management as well as transparency.[3] A third is Nigeria, which was the first country to 'make reporting of payments by all extractive companies and revenues received by government legally binding under national legislation' (EITI 2012b: 2). The extension of the principle of transparency occurs both by enacting a general model and meeting the demands of validation, and also through the imitation and modification of exemplary cases. As we might also say in Tardean terms, invention, a process of political and economic invention in this case, occurs along a pathway of imitative modification (Tarde 2001 [1890], Barry and Thrift 2007).

The narrowness and specificity of the Transparency Initiative suggests a number of responses. Harry West and Todd Sanders, for example, have ventured that 'in the globe's constituent localities, key words such as transparency, conveying notions of trust, openness and fairness, must dance endlessly across the same terrain as vernacular key words expressing suspicion, hiddenness, and treachery' (West and Sanders 2003: 12). Certainly in Azerbaijan it is easy enough to hear stories of oil revenues actually spent by the government. 'Public secrets', which may or may not be true, are unsurprisingly common. Ethnographic studies of post-socialist societies have provided rich accounts of the multiple forms of networking and favours in post-Soviet economic life (Ledeneva 1998, Humphrey 2003, Yalçin-Heckmann 2010). Historical and sociological studies of the Caucasus, in particular, have documented the failure of national governments to exercise control over local elites in Georgia and Azerbaijan during the Soviet era (Suny 1994). Georgi Derluguian argues that, following their appointment during the Soviet period, Heydar Aliev and Eduard Shevardnadze 'realized that their primary aim had to be to placate Moscow while consolidating their local power base by appointing local clients' (Derluguian 2005: 201, cf. Hibou 2004).

West and Sanders focus on the opposition and relation between transparency and suspicion (or openness and conspiracy). This is, however, only part of the story of transparency. Their essay concentrates on the production and circulation of rumour and suspicion, demonstrating in their words 'how tenous, even illusive, trust is ... in the midst of the turbulent transformations defining post-socialist societies' (West and Sanders 2003: 11). While this may be true in general, what is striking about the Transparency Initiative is not that its operation necessarily conceals anything, but that it is not expected to reveal much. It does not necessarily hide the truth, but leaves a huge amount unsaid. It allows a vast space of discretion: the realm of what one chooses not to know, does not investigate, or deliberately overlooks. On one occasion before travelling to the Caucasus, I met with some officials working for the Department for International Development in London. As I left the building one of them offered me some words of wisdom: you have to *avoid* listening to rumour. His advice is one working definition of how to exercise discretion, a necessary feature of transparency.

But if transparency is rendered so specific in its focus, it is, therefore, also achievable. One official noted that the Azerbaijan government found little difficulty in signing up to the Transparency Initiative because the Oil Fund[4] was *already* transparent and so did not require any substantial changes in the way in which it operated.[5] Since then, Azerbaijan, in conjunction with BP and other companies, has taken a leading role in the Transparency Initiative. It was the first country to be validated – and therefore publicly recognized – as transparent by the time of the biannual EITI conference at Doha in February 2009. According to a different assessment, the corruption perception index produced by Transparency International, Azerbaijan was 143rd in the same year (Transparency International 2009). It is quite possible that Azerbaijan, when its performance is assessed in different ways, can both lead the development of practices of transparency and yet, at the same time, not be transparent at all (cf. Guliyev and Akhrarkhodjaeva 2009).

Assembling Civil Society

In any case, the primary issue for the Oil Fund was not how to produce an account of the payments received from international oil companies, which was easy enough, but how to produce the right kind of witnesses, including stakeholders, oil companies, DFID and NGOs who might accept these claims as true.[6] This problem was also central to the preoccupations of EITI and its NGO supporters, including the Open Society Institute:

> Accountability, transparency, and public oversight require the creation of checks and balances and a separation of powers among an array of institutions established to oversee the overflow of oil and natural gas revenues. They also *demand input from civil society* and the creation of a powerful sense of public ownership of the revenues ... the chances for success of these funds would be improved by strengthening parliamentary oversight, improving budgetary transparency, and *establishing independent citizens' advisory councils to raise public awareness about and conduct monitoring* of the countries' oil and gas revenues. (Caspian Revenue Watch 2003: 6, emphases added)

Viewed in this way, what was central to the Transparency Initiative was not just the publication of information, but the progressive collection of persons who would be able to have input into, monitor and exercise oversight over the transparency process (see also World Bank 2008: 77–78). It implied the creation of a triangular relationship between government, the oil companies – both of whom published information – and NGOs, who exercised oversight over what information was published. One Azeri NGO informant expressed the problem in terms of the need to create a space within which public discussion was possible and 'information sharing' could take place.[7] The problem in Azerbaijan was that NGOs did not necessarily behave in this way. A few took oppositional positions, rejecting the existence of oil industry developments that were already in process. Other NGOs, conversely, were said to be more or less directly associated with the government (cf. Wilson 2005). Some were cautious about what they said in public. A few sought to uncover, through their own investigations, what they believed to be the true story of the ways villagers had been deceived by the government, for example by changing land ownership records. Such investigations went far beyond the limited demands of the Transparency Initiative. The difficulty that those supporting the Initiative confronted was how to foster an appropriate form of critical engagement with the problem of how to measure and manage oil revenues rather than, for example, to confront the government directly. In effect, 'civil society' had to emerge in a form through which it could perform the specific role expected of it.

Moreover, during this period, opposition to the Azerbaijan government was weak. Although reports from the Organisation for Security and Cooperation in Europe (OSCE) Election Observation Missions reported that elections in Azerbaijan failed to meet OSCE standards for democratic elections (OSCE 2003), Western observers argued that there was no credible alternative to the government of Heydar Aliev (see also Cheterian 2010). In Georgia, by contrast, accounts of electoral fraud in 2003 had played a critical role in the Rose Revolution that led to the fall of the Shevardnadze government (L. Mitchell 2009: 58–62, Companjen 2010).[8] Unlike in Azerbaijan, Western governments considered Georgian civil society to be already sufficiently developed for a change of government to be possible. In these circumstances, the principle of transparency could be applied not just to the apparatus of oil revenue payments, as it was in Azerbaijan, but to the apparatus of elections (cf. Coles 2004).[9]

The behaviour, freedom and capacity of civil society are considered critical to the success of the Transparency Initiative. In 2006 it was agreed that countries that had signed up to EITI would, in general, need to be 'validated' by 2010. To this end, what was termed 'a validation grid' was agreed upon, against which countries would be assessed. This grid set out eighteen validation criteria including, for example: #2 'has the government committed to work with civil society and companies on implementation'; #5 'has the government established a multi-stakeholder group to oversee EITI implementation'; #6 'is civil society engaged with this process'; #8 'did the government remove any obstacles to EITI implementation'; #13 'has the government ensured that government reports are based on audited accounts to international standards' (EITI 2006). Each of these grid indicators was expanded on further so that, for example, 'civil society groups involved in EITI should be operationally, and in policy terms, independent

Figure 3.1 Making Politics Transparent: public meeting held prior to the Georgian Parliamentary Elections, Tbilisi, March 2004. Photo taken by author

of government or the private sector' (ibid.: 14). The 'validation grid' additionally addressed the need for 'outreach by the multi-stakeholder group to wider civil society groups ... including coalitions (e.g. a local Publish What You Pay Coalition), informing them of the government's commitment to implement EITI, and the central role of companies and civil society'. On the basis of these criteria, Azerbaijan was judged to have met the requirements for validation. However, the auditors noted that the government had not fully established a permanent multi-stakeholder group (#5):

> In taking a view regarding Azerbaijan's compliance with this indicator, we have considered the wording of this IAT [Independent Assessment Tool] very carefully and considered the historical context of the EITI process in Azerbaijan. We believe that whilst Azerbaijan's previous institutional structure for EITI implementation enabled the achievement of the EITI's key objective, namely the regular publication of EITI reports without unexplained discrepancies, the formation of a permanent MSG [Multi-stakeholder Group] will enable stronger multi stake holder engagement in overseeing the strategic development of the EITI in Azerbaijan. (Coffey International Development 2009: A1-1, Crude Accountability 2012: 14)

Three observations follow from this. First, the validation process makes clear that the Transparency Initiative is expected to be performative. It is intended to foster the creation of the kind of civil society or public sphere imagined by Western social and political theorists such as David Held and Mary Kaldor (e.g. Held and Koenig-Archibugi 2005). Writing on the Azerbaijan-Armenian conflict over Nagorno Karabakh, Kaldor argued that civil society and transparency provided the solution to the problem of conflict resolution: '[international organizations] could strengthen civil society to a much greater degree in the peace process so as to stimulate public discussion and mobilize greater public support' (Kaldor 2007: 179). A similar logic is embodied in EITI. It is a device intended to foster the formation of a civil society prepared to engage in public discussion. According to a high-level panel set up to advise BP on the development of Caspian oil, BP's transparency could have an 'important and positive impact' on the 'free exchange of ideas' in Georgia and Azerbaijan (BTC/CDAP 2003: 13). In this context, transparency turned out to be a political device as well as an economic one. Kaldor and Held's notion of civil society derived not from natural resource economics but, at least in part, from relations with civil society organisations in Eastern Europe in the 1970s and 1980s and the theoretical resources of post-Frankfurt school critical theory (Habermas 1990).

Secondly, although it is concerned with a wide set of issues such as civil society engagement, and the existence of government 'obstacles', the validation process addresses these issues in a particular way. Validation was conducted by auditors whose names were chosen by the Azerbaijan government from a list drawn up by the EITI secretariat and board, which did not include any Azerbaijani nationals. In this way, the validation process would not be contaminated by too much knowledge of local complexities, prejudices or conflicts of interest. In Azerbaijan I carried out research in conjunction with an anthropologist who already had conducted her own ethnographic fieldwork in Azerbaijan as well as a number of interviews with NGOs interested in the development of the oil industry. In these circumstances, I immediately gathered the excess of detail that is typical of fieldwork. Once, after a long interview with a member of an NGO, my colleague criticised me for not asking more probing and critical questions about the personal connections and relations of our NGO informant. For the anthropologist, what was external and hidden (the 'realm of suspicion' to use West and Sanders' term) was of greater interest than what was in the open and merely presented to us. What was performed for our benefit pointed to the existence of social relations beneath (Strathern 2000). Our research highlighted the existence of things that were difficult to make public (Quattrone 2006).

Thirdly, if the validation process has a quasi-Habermasian logic, seeking to forge a space within which rational debate concerning matters of public interest can occur, the existence of that space is confirmed in a particular, technical way. In conducting the validation process quickly and without substantial local knowledge, validation will leave much unknown, and therefore unreportable. Seen in this context, the audited accounts produced by the Transparency Initiative are arguably less important than the political assembly that needs to exist for these figures to be examined. The logic of the Transparency Initiative, for many of those associated with it, is that this political assembly or civil society forum will progressively learn to ask the right questions, gradually demanding more information about other matters beyond the narrow remit of EITI itself (concerning, for example, production-sharing agreements, concessions, taxation, bonuses and so on). The technicality of the issues that civil society needs to be concerned with are important, partly because they are technicalities and civil society, if it is to properly function as such, needs to concern itself with technicalities (Schulz 2005). There is an echo here of an argument put forward by the late nineteenth-century French sociologist, Gabriel Tarde. In *L'Opinion et la foule*, Tarde looks forward to the day when the public would read and digest social statistics rather than indulge in the highly contagious imitative and affective forms of behaviour characteristic of street demonstrations 'irresistibly drawn by a force with no counterbalance' (Tarde 2006 [1901]: 16). The Transparency Initiative embodies this political logic. It is expected to provide a technical solution to the management of affect, a preventative cure to the contagious forms of imitative behaviour that Tarde saw in the late nineteenth-century urban crowd (Salmon 2005, Toscano 2007, Borch 2012). Transparency, in effect, is a device intended to foster the formation of a rational civil society and a rational government, albeit in embryonic form. Conflict resolution between government and civil society is expected to occur not by opening up a potentially uncontrollable space of antagonism resulting in the kinds of violent clashes between police and demonstrators that have occurred on the streets of Baku, but by focusing protagonists on the mundane problem of how to generate and verify particular matters of fact (Shapin and Schaffer 1985). In effect, transparency both addresses an ongoing political situation and is intended to contain and manage it in a particular form. Debate should begin, in this view, through the examination of particular and limited details, not with wider demands for social justice and the redistribution of wealth (Rancière 1998, 2006). The enactment of revenue transparency would not reveal that much in the short term, but it would be the basis, it was claimed, for a different political future.

A Public Experiment

Talk of the importance of transparency has certainly become pervasive, not least in relation to discussions of the extractive industries. Yet it would be a mistake to equate transparency, as Christopher Hood suggests, with an all-encompassing regime of surveillance (Hood 2006: 8–9). And Stephen Collier argues persuasively that the same could be said of neo-liberal economic reform programmes in the post-Soviet period: 'The tools of the new economics of regulation were invented precisely as a new form of critical visibility through which intransigent things, embedded norms, and patterns of social provisioning could be brought into view, down to minute technical details, as the product of a prior governmentality that had to be rationalised. And this rationalisation is designed to take shape precisely through the selective and in some cases quite *limited* deployment of … microeconomic devices' (Collier 2011: 242, emphasis added). Collier's observation could also be made about the principle of revenue transparency; indeed what is striking about the operation of EITI is that it appears to reveal rather little.

Yet if we focus not on what the implementation of revenue transparency is expected to reveal but on what it is expected to perform or do, then this remark becomes less of a paradox. Three observations follow. First, it is important to recognise that the implementation of transparency in the extractive industries takes an evolving and experimental form. As an experiment, revenue transparency is intended not only to effect a progressive transformation in the world within which it is conducted, but also to persuade others that the results of this experiment are both true and valuable. It is precisely in its lack of transparency, as we have seen, that Azerbaijan has provided a particularly suitable location for the experimental application of transparency in practice. Secondly, and at the same time, oil turns out to be not just another industry whose operations can and should become transparent, but one that is perceived to be acutely in need of or receptive to transparency; for it is the lack of transparency of the oil economy that is thought by some economists to be at the root of the economic problems of many oil-producing states. Thirdly, transparency implies not only the publication of specific information but also the formation of a society, a public, that is in a position to recognise and to assess the value of – and if necessary to modify – the information that is made public. The operation of transparency is addressed to local as well as global witnesses, yet these local witnesses are expected to be properly assembled, and their presence validated. There is thus a circular relation between the constitution of political assemblies and accounts of the oil economy: one brings the other into being (Mitchell 2011). Extractive industry transparency is not just intended to make information public, but to govern the constitution of a public that is interested in being informed (see Chapter 5).

While there has been a great deal of interest, especially on the part of economists, in the potential value of transparency as a solution to the problem of the resource curse, the significance of transparency for any account of the contemporary politics of oil is much wider than this. In part the progressively increasing desire for transparency has been driven by the conjunction of intersecting and competing demands made by international organisations, institutional investors (Clark and Hebb 2005) and multinational companies, as well as civil society organisations such as Revenue Watch and the Open Society Institute (cf. Djelic and Quack 2010). In part it has been governed by the enactment of a growing range of national and international principles, conventions and regulations, as well as national legislation (Abbott and Snidal 2000, Larner and Walters 2004, Agrawal 2005, Fairhead and Leach 2003, Djelic and Sahlin-Anderson 2006, Jessop 2008). A number of international agreements and principles that demand certain degrees of transparency have come to have special importance to the politics of oil. In addition to the EITI (2003), they include the Åarhus convention on public information disclosure (1998), the OECD Guidelines on Multinational Enterprises (2000),[10] the Equator Principles (2003)[11] and the UN Declaration on the Rights of Indigenous Peoples (UN 2007). Moreover, projects supported by international financial institutions such as the World Bank and the European Bank for Reconstruction and Development (EBRD) are expected to conform to the operating principles and guidelines of these institutions, which also enshrine the principle of transparency. This evolving nexus of guidelines, principles, agreements, laws and codes of practice constitute an overlapping and uneven series that both modify and supplement existing national legislation, and that in practice may either contradict or complement each other.[12]

Having explored the operation of EITI in relation to the specific question of revenue transparency, from Chapter 5 onwards I turn to consider how the development of the BTC pipeline was subject to a much wider set of expectations for transparency. The contrast between the narrow range of information made public through the development EITI and the vast quantity of information made public in the period prior to and during the construction of the BTC pipeline was remarkable. The BTC pipeline, as we shall see, was to be materially invisible, but the invisibility of the pipe itself, following construction, co-existed with its informational visibility. Whereas the political experiment of revenue transparency took place under carefully controlled conditions in Azerbaijan, the political experiment of BTC was, in comparison, conducted across a more heterogeneous and less well regulated space, stretching across three countries.

Before examining the transparency of the BTC project in greater detail in later chapters, I turn in Chapter 4 to consider a further dimension of its politics. This revolves around the broader concern, which flourished from the mid-1990s on, with the ethical conduct of the oil industry, and especially its social and environmental responsibility. I highlight, in particular, how social and environmental responsibility came to be understood during this period as something that had to be demonstrated in public in order to be guaranteed. At the same time, I contrast these performances with the work of radical critics who focused their attention on what they took to be specific instances of unethical or irresponsible conduct on the part of the industry, while expanding the politics onto a larger plane.

Chapter Four
Ethical Performances

Critical observers might regard what we might call the ethicalisation of the oil business as a type of anti-politics, a symptom of what Slavoj Zizek has termed 'post-politics', where the possibility of ideological conflict has been displaced by a concern with good governance and collaboration between 'enlightened specialists' (Zizek 2004: 72). The growing interest in ethics and corporate responsibility has indeed derived in part from a desire to manage the domain of the political. Yet to view the ethicalisation of the oil business as merely a symptom of a post-political world is misleading. For, in practice, the ethicalisation of oil has not led to any straightforward reduction in conflict over the conduct of the industry. Indeed, critics of the industry have increasingly couched their criticisms of corporations in the language of ethics, highlighting the failure of oil corporations to live up to the standards that they profess and against which they should be measured. In these circumstances, in criticising the perceived immorality of the oil industry and its putative lack of genuine concern with issues of social justice, human rights and environmental sustainability, civil society critics add their own assessments of the industry's ethical performance to those already produced by the businesses themselves (Watts 2005, Gouldson and Bebbington 2007: 9). The result is a multiplication of accounts, assessments and reports about, as well as critiques and exposés of, the ethical conduct and responsibilities of the oil industry. The effect is what we might call, following Gramsci, a 'war of position' over the ethical conduct of the oil industry (Gramsci 1971).

Material Politics: Disputes Along the Pipeline, First Edition. Andrew Barry.
© 2013 John Wiley & Sons, Ltd. Published 2013 by John Wiley & Sons, Ltd.

The importance of ethics to the politics of oil raises three questions. One is the question of how the ethical or unethical conduct of the oil industry can be demonstrated or contested in practice (Dolan and Rajak 2011). For if the struggle between 'right and wrong' has become central to the politics of oil, how is it possible to demonstrate that something *is* ethical or unethical? The first part of this chapter considers how the ethicalisation of the oil industry is expected to be brought about through processes of monitoring and reporting on the existence of ethical and unethical conduct. In this way, the ethical actions of the industry are made explicit, whether in the form of published audits, environmental impact assessments, company reports or marketing. Lenders and regulators place demands on businesses not only to assess and mitigate the environmental, economic and social consequences of their activities and to respect human rights, but also to enter into a dialogue with those who are affected by the oil industry, and to provide evidence of these actions. In short, ethical conduct is considered to be guaranteed and demonstrated by being accounted for (see Osborne 1993, Born 2005a). This logic powerfully informed the development of the BTC pipeline and the particular manner in which its construction became politicised; and one aim of this chapter is to contrast this approach with the classical proposition put forward by the sociologist Emile Durkheim that the ethical conduct of business is best cultivated through the ethical norms of the professions.

A second question concerns the relations between the ethical conduct of the oil industry and both its material infrastructure and its products. The philosophical study of ethics has focused, in general, on the actions, virtues and responsibilities of human subjects and human interactions, and the ways in which modes of ethical conduct have been understood, fostered and policed historically (MacIntyre 1981). Yet the growth of concern with the social and environmental responsibilities of the oil industry is not only expected to affect human conduct and the relations between persons, but also the properties and agency of materials, and the accidents and events to which they give rise. This is not surprising, for materials are not just the stage on which to play out their relationships; social agency derives from assemblages of persons and things (Deleuze 1988, Pickering 1995). But if this is the case, in what ways has the ethicalisation of the oil industry been associated not only with changes in the conduct of persons but also with changes in how the properties and behaviour of materials are monitored, measured and managed?

A third and related question concerns the relation between accounts of specific events, including accidents and acts of violence, and general claims about the conduct of either particular corporations or states or the oil industry as a whole. These relations between specific events and widespread forms of behaviour are formulated by different institutions in different ways. Industry accounts of corporate social responsibility and transparency stress the ways in which ethical conduct has become routinised and embedded in the practice of the corporation. At the same time, the industry's actions are expected to conform to a growing range of guidelines, standards and voluntary codes of conduct that embody ethical principles and that are, in principle, globally applicable (Thompson 2012). A regulatory review of environmental and social issues on the BTC pipeline notes, for example, that BP is committed to the UN Universal Declaration of Human Rights, the 1977 International Labour Organisation 'Tripartite Declaration of Principles Concerning Multinational Enterprises and Social Policy' and the 1976 OECD 'Guidelines for Multinational Enterprises', as well as a series of company 'policy expectations' including 'not to offer or accept bribes', to 'respect the law in the countries and communities' in which BP operates and to 'evaluate the likely impact of our presence and activities' before making any major investments in a new area (BTC/ESIA 2002h: 3). In contrast, reports of the unethical conduct of the oil industry often focus on particular cases of, for example, actual or potential pollution or specific instances of violence or of the abuse of human rights, or the failure to evaluate possible impacts.

Critical accounts of specific events are often intended to raise questions about the unethical conduct of individual oil companies or of the oil industry more broadly. The particular case is taken to be exemplary of a general problem, an element of what I have termed a political situation that transcends the specificity of the case. Critics may be concerned with the question of whether the unethical or ethical conduct of oil companies can be demonstrated in general, establishing a context that frames the act as politically significant (cf. MacIntyre 1981: 9). The logic of such demonstrations is, I will suggest, abductive. An abductive inference is not based on statistics or legal judgement, or on an understanding of the dynamics of an economic system, but on the force of the example. Abduction, as C.S. Peirce explained, gives us reasons to suspect something is true, even if we do not know it for certain (Peirce 1934). In the second part of the chapter I focus briefly on a historical example of the politics of abductive inference: the controversy surrounding the decommissioning and disposal of the oil storage facility, the Brent Spar. The case has already been widely discussed, but it is examined anew here because it has been taken as a formative historical event in the development of an ethical orientation on the part of the oil industry,[1] which preceded the development of the BTC pipeline.

I have already discussed how a number of non-governmental organisations, including the World Wildlife Fund, Amnesty International and the Baku-Ceyhan Campaign developed various criticisms of the BTC project, focusing their attention on the ways in which the construction of BTC systematically failed to conform to a series of guidelines and standards proposed by the international financial institutions and set down in a range of international agreements (Amnesty International 2003, Baku-Ceyhan

Campaign 2003a&b). In the final part of this chapter, however, I highlight the interest of writers, artists and film-makers in the construction of the pipeline and the political practices to which this led, interventions that co-existed with the quasi-legal political practices of international and local NGOs. In particular, I consider the work of Platform, a small but influential group of artists and researchers based in London concerned with the operation of the oil industry, which played a pivotal role in the Baku-Ceyhan Campaign. Platform, I suggest, sought to disrupt the formation of an infrastructure – an oil pipeline – that was, they surmised correctly, informational as much as it was material. Here, I contrast Platform's approach with other types of political engagement with the work of corporations in general, and the oil industry in particular. Social theorists and philosophers from Max Weber onwards have long been anxious that the growing importance of science and technology would come to dominate political life, leading to what Habermas once termed 'the scientisation of politics and public opinion' (Habermas 1971, cf. Eden 1999). This criticism is an important one. But the ethicalisation of the oil industry, I will suggest, has not led to the 'scientisation of politics' nor to the development of a post-political world. On the contrary, it has led to the politicisation of the industry's ethical conduct.

Ethics and Business

In a series of lectures on 'The Nature of Morals and Rights', originally delivered at the University of Bordeaux in the 1890s, Durkheim explored the problem of what he called 'moral and juridical facts'. In accordance with the general principles set down in his *Rules on Sociological Method*, he understood these to be the 'rules of conduct that have received sanction' (Durkheim 1957: 1). Moral and juridical facts, Durkheim argued, lead to sanctions, yet such sanctions do not follow from 'the act taken in isolation but from the conforming or not conforming to the rule of conduct already laid down' (ibid.: 2). In order to understand what sanctions exist in a particular society, one needed the insights of comparative history and ethnography.

Although Durkheim had an interest in moral facts in general, he argued that there was also a realm of moral facts that had force within specific groups and, in particular, the professions. Professions were distinctive, Durkheim argued, because their moral order was not affected to any great degree by public opinion. At the same time, the professions had established particularly strong systems of morals, which operated within but not outside of the professions themselves. As a result, moral rules were localised and polymorphous: 'whilst public opinion, which lies at the base of common morality, is diffused throughout society, without our being able to say exactly that it lies in one place rather than another, the ethics of each profession are localized within a limited region' (ibid.: 7). Moral rules were particularly strong, he argued, in those professions that had close relations to the activities of the state (for example, education, the army and the law).

Given this analysis, Durkheim was troubled by the growing importance of business in society. In his view, business was progressively displacing older institutions of moral authority, including the Church, and even undermining new institutions of authority, such as science (cf. Osborne 1993, du Gay 2000, Mol 2006). The weakness of corporate bodies as well as the lack of stable and close connections between individuals involved in business gave economic life 'an amoral character'. In his bleak analysis, the manufacturer, the merchant and the employee were all 'subject to no moral discipline whatever' (Durkheim 1957: 12). Durkheim's solution to this 'evil' was to call for the different groups involved in business to develop stronger professional ethics and, by implication, to apply stronger sanctions to those who transgressed the moral rules.

Although it is now commonplace to talk about the ethical conduct and corporate responsibility of businesses, the moral system of contemporary business has not followed the logic of Durkheim's proposal. Where Durkheim argued that professional ethics were not affected by public opinion, contemporary efforts to render business more ethical are strikingly bound up with public opinion. If, according to Durkheim, the public had little interest in matters of professional ethics, contemporary businesses are particularly concerned with the public's perception of their ethical performance. Moreover, while Durkheim stressed the need for professionals working within business to become more ethical, contemporary businesses often draw on the expertise of external experts in matters such as reputational risk management, environmental audit, social impact assessment and community development; indeed a whole spate of businesses have arisen devoted to such institutional audits. There has been an effort by businesses both to render their activities ethical and to demonstrate to others that this has been achieved, and not by relying on the ethical conduct of professionals. If Durkheim thought that those who worked in business needed to turn inwards, to ensure that employees conduct themselves in an ethical manner, in contemporary businesses the solution to the problem of ethics appears to be conceived in terms of turning outwards (Shultz et al. 2000, Thrift 2005).

A common critical response to the growing concern with ethical conduct and the social and environmental consequences of the oil industry is an empirical one. How much, in reality, do individual oil companies live up to the claims that they make about themselves? Are businesses actually as committed to matters such as environmental sustainability and community investment as they make out? Although this response may lead to criticisms of the *practices* of particular businesses, they are not necessarily critical of the *form* in which businesses seek to foster their own ethical conduct. Indeed, the reverse may be true. Part of the logic of recent efforts by businesses to render themselves more ethical is for them to invite and anticipate potential criticism, in this way making use of the critical resources of observers, including expert consultants, stakeholders, public opinion and civil society. In seeking to become more ethical, the problem confronting contemporary businesses is not to exclude criticism, but rather to manage, channel and translate what we might call the 'immaterial labour' of criticism (Lazzarato 1996) – both by recognising its value, and by turning it productively into a source of value, as reputation (Gouldson and Bebbington 2007: 9, Fombrun and Rindova 2000, Clark and Hebb 2005, Holzer 2010). In these circumstances, the range of practices involved in generating and accounting for the ethical performance of the oil industry is now considerable. They include not only social and environmental impact assessment and community investment, but also marketing, public relations, reputational risk management, stakeholder engagement and public engagement. Many of these activities are contracted out to specialist firms, consultants, NGOs or universities dispersed across both oil-producing and oil-consuming states. As we shall see in Chapter 5, the development of the BTC pipeline project entailed a vast exercise in stakeholder engagement and public consultation, the results of which were also made public. Yet, in turn, international NGOs claimed that the public had not been consulted sufficiently or properly and that criticisms had not been sufficiently addressed. Metropolitan intellectuals and activists did not challenge the importance of public consultation, then, but sought to reveal its limitations, claiming that they themselves were better able to represent the interests of communities in distant locations.

As Michael Power (2007a) argues, there is a weakness in the idea that ethical conduct is best improved through communication of its performance to others: the danger is that businesses become preoccupied with reporting requirements, such that they behave ethically primarily for the sake of these reports. Indeed, the interests of business in demonstrating ethical conduct – by meeting standards and guidelines and making its commitment to corporate social responsibility public – may lead it to neglect those aspects of its conduct that cannot easily be measured or publicised, or that are not the subject of demands and expectations on the part of regulators, the legal system or even critics. At the same time, published accounts of ethical conduct may function as a way of defending businesses against criticism; while critics, in turn, can become preoccupied with the ethical performance of business as it comes to be defined by international standards. These observations suggest that we should not focus solely on the question of whether or not businesses act ethically according to accepted measures and guidelines. Rather, we also need to consider how and why accounts of the ethical conduct of business are produced and contested, and with what consequences (Power 2007a&b). In the remainder of this chapter I turn to the question of how critics have contested the claims made by the oil industry both about its ethical concerns for the environment, society and human rights, and about its conduct in practice.

Ethics, Politics and the Logic of Abduction

If, as Chantal Mouffe argues, the 'struggle between left and right' has been displaced by one between 'right and wrong' (Mouffe 2005a: 5), then those engaged in this struggle have developed a series of ways of demonstrating that an action or event is ethical or unethical. On the one hand, the growing body of law and guidelines governing the operations of the oil industry itself provides the basis on which controversies can emerge. Indeed, disputes over the construction of the BTC pipeline were frequently articulated in relation to international guidelines, not necessarily directly addressing the impact of the pipeline itself, but rather challenging whether BP and its consultants had properly assessed the environmental and social impact of the pipeline or had consulted or compensated affected communities. In subsequent chapters I examine some of these disputes in more detail. On the other hand, critics of the oil industry have also sought to demonstrate that individual objects and events can be taken as indicators of the ethical conduct of the industry in general and should, in this sense, be understood as more than merely individual issues or legal cases. In this context, what matters is not just the specificity of the disputed issue, but the way in which the issue reveals the existence of more widespread and problematic tendencies in the way the oil industry operates.

Certainly, the history of the oil industry includes a number of highly politicised accidents and events that have been taken to be manifestations of wider failings and deeper causes. These include the environmental disasters that followed from the massive oil spills from the Torrey Canyon (1967) (Sheail 2007) and Exxon Valdez (1989) (Gramling and Freudenberg 1992) tankers, and the Deepwater Horizon rig (2010), the loss of life following the fire on Piper Alpha platform in the North Sea (1988) (Woolfson et al. 1996), and the widespread pollution of the Niger delta (Watts 2008). In many of the disputes that developed around these disasters and accidents, the generation and circulation of information, and the contestation of claims to knowledge about what happened and why it happened, came to be of critical importance. However, in considering the question of the political significance of individual accidents and events, I focus briefly here on the controversy that erupted over the decommissioning of the North Sea oil platform, the Brent Spar, in 1995. The case of the Brent Spar is an instructive one precisely because it was not associated with a particular accident, nor with violence or the occurrence of pollution; yet it was nonetheless taken by critics to be an expression of the unethical conduct of the oil industry. It was a controversy that, like the BTC pipeline, revolved around competing factual claims about the construction of a major piece of infrastructure. In these respects, there are clear links between the controversy surrounding the Brent Spar and those that arose around the BTC pipeline nearly a decade later. At the same time, the Brent Spar controversy helped to generate wider concerns about the importance of corporate social responsibility, and these in turn formed part of the political situation within which the BTC controversies subsequently developed.

The broad outlines of the case of the Brent Spar are well known. On 30 April 1995, activists from Greenpeace occupied the 14,500 tonne Brent Spar oil storage facility in the North Sea, protesting against the proposal by Royal Dutch/Shell to dispose of the facility at sea.[2] The oil company's proposal had previously been submitted as the Best Possible Environmental Option (BPEO) and approved by the UK's Department of Trade and Industry. However, Greenpeace claimed that the platform contained large quantities of toxic sludge, a contention that was denied by Shell. Nonetheless, the environmentalists were able to mobilise public support against Shell, particularly in Germany, gaining additional support from Chancellor Helmut Kohl. Responding to pressure, and despite continuing support from the UK government (including the intervention of Prime Minister John Major), Shell withdrew its proposal in June 1995 (Dickson and McCulloch 1996). Subsequently, the Brent Spar was temporarily moored in Erfjord in Norway, awaiting a decision regarding its future. After the oil company had solicited further proposals for its disposal, and after it had engaged in public consultation, it was decided that the Brent Spar would provide the base for a permanent quay at Mekjarvik, near Stavanger. On 10 August 1999, Shell announced 'a symbolic end to its deconstructing activities' (Shell nd).

While the controversy over the decommissioning and deconstruction of the Brent Spar was resolved many years ago, the case has been cited copiously in literature on environmental politics, corporate ethics, public relations, decommissioning in the oil and gas industries, science and technology studies, risk communication, and reputational and brand management (Dickson and McCulloch 1996, Löfstedt and Renn 1997, Gorman and Neilsen 1997, Bennie 1998, Rice and Owen 1999, Huxham and Sumner 1999, Smith 2000, Gordon 2001, 2002, Livesey 2001, Yearley 2005, Power 2007a, Holzer 2010). The political scientist Maarten Hajer cites the Brent Spar as one of five examples that illustrate 'challenges to the classical-modernist way of policy making and politics in a changing world' (Hajer 2003: 177, Holzer 2010, chapter 3). As a result of the Brent Spar, Michael Power suggests, 1995 was 'a critical year for the emergence and intensification of managerial reputation' (Power 2007a: 128). The case has also been considered catalytic in leading to a movement on the part of

Shell as a corporation 'from a taken-for-granted discourse of economic development towards a cautious adoption of the language of sustainable development, which attempts to balance interests of economic development with environmental well-being' (Livesey 2001: 59). For the director of public affairs at Shell UK, in the wake of the Brent Spar controversy 'businesses will ... have to come to grips with an area of deep seated emotions, subconscious instincts and symbolic gestures' (cited in Regester and Larkin 2005: 94). Along with a handful of other major events and accidents in the history of the oil industry, the Brent Spar has come to play a critical part in the history both of environmental politics and of the ways in which the oil industry engages with social and environmental issues. Thus, 'Brent Spar was not just a decision about a reasonably big engineering project, or even important as a precedent for several hundred oil industry decisions about disposal, but was the pivot on which a more general business re-appraisal of the environment took place' (Jordan 2001: 8).

The view that the Brent Spar controversy was an historical event, a catalyst or pivot, has been performative: subsequent accounts of its implications have reinforced its historical significance. Even if there is no consensus about what actually happened, nor about its continuing significance today, there is no doubt that something of consequence happened. The Brent Spar controversy demanded attention. But while commentators today recognise the significance of the occupation of the Brent Spar, the question of the general significance of the case, and of the relation between particular facts and general claims, was central to the controversy all along. Whether the Brent Spar was destined to play a key part in an emerging political situation with future ramifications, or was simply an isolated and time-limited controversy, was therefore itself at issue. According to the UK government, the Brent Spar had no general significance at all. As a junior minister noted in a discussion in a UK House of Lords select committee, 'clearly, what is required will be decided on a case-by-case basis'. Indeed, he considered this an appropriately scientific approach given that the toxic sludge had 'a very small proportion of heavy metals which are not significantly different from those found in a similar weight of plankton' (House of Lords 1995: 1538). In this view, the Brent Spar was simply a complex, multi-dimensional technical object, and few wider conclusions, scientific or otherwise, could be drawn from the case.

For Greenpeace, in contrast, the case of the Brent Spar was important precisely in so far as it was an index of general problems including, notably, the practice of sea dumping. The Brent Spar was not particularly significant because of its specificity, but because it revealed the existence of an as yet unrecognised larger issue. Indeed for the campaign director of Greenpeace, Chris Rose, the success of their campaign was demonstrated by the UK government's imperceptible shift away from the 'situation where "case-by-case" actually meant dumping' towards a policy of 'anti-dumping' (Rose

1998: 52). Following the election of a Labour government in 1997, the controls were further tightened by an assumption that sea dumping should not occur unless a special case could be made (ibid.: 72). At the same time, the oil industry came to recognise the importance both of its responsibility for decommissioning and, more broadly, of the relation between environmental performance, corporate social responsibility and reputational risk.

How can an assemblage such as the Brent Spar come to matter politically and to have such collective significance? The logic of Greenpeace's argument took the form of an abductive inference. By this I mean, following Peirce, 'a variety of nondemonstrative inference, based on the logical fallacy of affirming the antecedent from the consequent ("if p then q; but q, therefore p"). Given true premises, [an abductive inference] yields conclusions that are not necessarily true' (Boyer 1994: 147, see also Gell 1998). Peirce, however, preferred to define an abductive inference even more precisely, with reference to specific conditions under which it could lead to a logical conclusion:

> The form [of the abductive] inference therefore is this:
>   The *surprising fact*, C, is observed;
>   But if A were true, C would be a matter of course.
>   Hence there is reason to suspect that A is true.
> Thus, A cannot be abductively inferred, or if you prefer the expansion, cannot by abductively conjectured until its entire content is already present in the premise, 'If A were true, C would be as a matter of course.' (Peirce 1934: 117, emphasis added)

Abduction is not simply a way of decoding an object or text (see Barthes 1973) and, I will contend, its operations exceed linguistic interpretation. Abduction can be understood as a form of inference that, rightly or wrongly, draws the addressee or audience towards the existence of an agency, or causal agencies, from which the object or action derives (Eco 1976, Gell 1998: 14). It both turns audiences towards and constitutes the existence of forces beyond the object or event itself. An observation of a smile in a visual image, for example, leads to the abduction that the person smiling is friendly, which may generate a response. The smile elicits effects that are more than merely cognitive, but affective (McCormack 2007, Thrift 2008). But the affective response in turn alters the circumstances within which abduction occurs.

Certainly, the Brent Spar demanded the attention of observers (cf. Gell 1999: 211). Once it was occupied, the Brent Spar elicited responses, and indeed that was the intention. These responses built upon particularly expressive or affect-inducing features of the occupied object: the presence of toxic sludge and oil, its location in the North Sea, its rust and decay, and the heroism of the Greenpeace activists who had scaled and occupied it (cf. Carter and McCormack 2006). The abductive inference could be drawn from all this, correctly or incorrectly, that Royal Dutch/Shell, the British government and the oil industry more generally were not exercising proper care for the sea, viewing it simply as place for waste disposal or large-scale corporate dumping, a backyard into which dangerous debris could be thrown. Once this abductive logic was catalysed, accepted or internalised, the case of the Brent Spar could not be decided on its individual merits; rather, it was taken to reveal a set of actors – governments, the oil industry – that might behave similarly in the future. Although both the British government and Shell argued that Greenpeace was mistaken on a number of points of fact, this did not diminish the effectiveness of Greenpeace's actions as a mode of political performance. Rather, through abductive inference, the occupied Brent Spar directed its audiences to the general significance of the event, undermining the claims made by Shell and government scientists that the facility had only to be considered in terms of its specificity; and indeed, for a time after the Greenpeace occupation, the Brent Spar – acting like a 'trap' – dominated political debate (Gell 1999, Born 2011).[3]

The case of the Brent Spar has therefore to be seen as more than a momentary political event, or an expression of a specific political context. For the occupation of the Brent Spar, and the inferences that were drawn from it, contributed to a larger transformation in the very political conditions within which the oil industry operated. As Michael Watts (2005: 9.21) observes, a broader concern with corporate social responsibility had already emerged in the 1980s and flourished in the 1990s. However, in the context of growing public disquiet about the links between oil production and climate change, pollution, corruption and violations of human rights, public interest in the ethical conduct specifically of the oil industry escalated further in the late 1990s and early 2000s (ibid.). In these conditions the occupied Brent Spar was not so much an individual case, but itself catalysed a political situation in which questions of environmental ethics would become critical to the governance and politics of oil. In 2000, BP rebranded itself as an ethical company that would move 'Beyond Petroleum'; while in 2001, Greenpeace launched a campaign against ESSO, claiming that it was the 'no. 1 global warming villain' (Greenpeace 2012). In the early 2000s the World Bank Group promoted 'pioneering work on environmental and social mitigation' along the route of the Chad-Cameroon oil pipeline (IFC nd), while in 2002, as we have seen, the Extractive Industries Transparency Initiative was launched. In this rapidly evolving and contested political situation, the oil business, its lenders, consultants, regulators and critics could not agree about whether the conduct of the industry was ethical or not; but the question of its ethical conduct nonetheless came to structure the manner in which the activities of the oil business was both governed and politicised.

Art, Research and Politics

There was no occupation of the BTC pipeline by international critics equivalent to the occupation of the Brent Spar by Greenpeace. Although, as we shall see, villagers engaged in actions that stopped construction work at various points along the pipeline route, particularly in Georgia, international environmental and human rights NGOs campaigning around the pipeline did not employ the techniques of direct action that Greenpeace had used nearly a decade earlier. Nonetheless, these critics of the pipeline project did engage in practices and performances that sought to stimulate abductive inferences about the agencies that lay behind the pipeline's construction. In this way, like Greenpeace, they directed the attention of interested publics towards the relations between the pipeline and the complex of firms, international organisations and individuals that lay behind its construction. In what follows I take the work of Platform, a group of activist researchers and artists funded by the Arts Council of England and others, to be indicative of such practices. Platform has not only explicitly taken a stance in opposition to oil corporations, but it played a major role in the campaign that developed around the construction of the BTC pipeline.

Platform is not the only group of artists concerned with the operations of the oil industry. Other such artists include the film-maker Alfredo Jaar, the photographer Edward Burtynsky and the audiovisual artist Ursula Biemann (Biemann 2005, Franke 2005, Burtynsky 2009, Jaar 2006). But in relation to the oil industry the work of Platform is distinctive in two respects. The first is its close attention to the multiple connections between the oil-producing regions and the city of London, where Platform itself has been based (Marriott 2005). Platform's research and artistic practices set out to reveal the associations between, on the one hand, sites of social and ecological injustice and, on the other, networks of institutions and buildings – commercial and international banks, government agencies, consultancies and oil company offices – across London and elsewhere (cf. Hawkins 2013). For example, through conducting a guided walking tour of these institutions in the city, Platform sought to demonstrate the presence of the oil industry in the everyday life of London, revealing the evolving historical connections between the capital, the oil industry and the British Empire. This is a model of the artist as an experimental and critical researcher of the neo-colonial economy (cf. Foster 1995, Bishop 2012). It is an aspect of Platform's work that was taken up by the geographer Doreen Massey, who endorsed Platform's approach to the politics of place and space, observing that a map they produced of the companies and institutions related to Shell's operations in Nigeria 'evoked the presence within this place [London] of impacts on others beyond' (Massey 2010: 205, cf. Massey 2005).

If Platform has sought to open up the question of the embeddedness of the oil industry in the city, espousing a place-based politics, a second distinctive quality of its practice has been to establish alliances with organisations involved in explicitly political modes of action in relation to the oil industry. For Platform as a group, art is political when it is associated with explicitly political goals: 'Platform works across disciplines for social and ecological justice. It combines the transformatory power of art with the tangible goals of campaigning, the rigour of in-depth research with the vision to promote alternative futures' (Platform 2006). It was against this background that Platform helped to form the Baku-Ceyhan Campaign, along with Friends of the Earth, The Corner House and the Kurdish Human Rights Project (Platform et al. 2003). For Platform, the political situation they associated with BTC went far beyond the specific issue of the construction of the pipeline and its impact on the environment and human rights. Rather, the pipeline had to be understood in relation to the critical historical role of BP in the economic and political life of London as a hub in the global oil economy. BTC mattered not just in itself, but because of how it could be used both to condense and to express a critical analysis of the political geography of the global city.

Platform's inventive political and artistic practice looks very different from Greenpeace's spectacular occupation of the Brent Spar. Yet in certain respects, there are similarities between Platform's practice and the Greenpeace occupation; for Platform attempted, in effect, to enter into and render visible the informational infrastructure of the BTC oil pipeline, pointing to what Platform saw as its weaknesses and the lack of environmental and social concern concealed by this infrastructure. Platform's practice was based on the recognition that the pipeline was always more than a physical infrastructure. For the oil company was engaged not only in constructing such a physical infrastructure on a huge scale, but one whose properties and impacts were routinely measured, monitored and demonstrated to others. In this context, Platform's guided tour of London took in the offices of Environmental Resources Management, the company that carried out the BTC environmental and social impact assessment. Moreover, by producing and commissioning their own analyses and counter-reports, Platform developed a rich account of the complex of institutions involved in the development of the pipeline in London and elsewhere (Platform et al. 2003). In this way, Platform sought to show how the pipeline must be understood as the product of this specific 'corporate colonial' system (Marriott and Muttit 2006).

These ambitions are clearly articulated in a recent book jointly authored by James Marriott and Mika Minio-Paluello, both from Platform, entitled *The Oil Road: Journeys from the Caspian Sea to the City of London* (Marriott and Minio-Paluello 2012). The book, which received favourable reviews in the *Financial Times* and the *Guardian*, documents the authors' journeys along the BTC pipeline and beyond, from the Caspian to the Mediterranean and thence to the refineries of Western Europe and the financial centres of the City of London. Their narrative is supplemented by the authors' investigations of the stories that lie behind various points along the route as they encountered them both in 2009 and in 2002–4, the period of Platform's earlier involvement in the fact-finding missions carried out by the Baku-Ceyhan Campaign. In this way the authors explore 'the reality of this infrastructure [the BTC pipeline] and how it continues to be contested years after oil began to be pumped through it' (ibid.: 7). Where BP and the international financial institutions underlined the importance of transparency and corporate social responsibility to the BTC project, the authors dwell on the high levels of security and secrecy they encountered along the route. They observe what they experience as the intense security surrounding the Tbilisi offices of BTC and describe the oil fields of the Caspian Sea as a 'forbidden zone'. They document, with subtlety and care, the complex and uncertain links between the pipeline and the Georgian-Russian war of August 2008, tracking down the location of bomb craters near the route of the smaller Baku-Supsa pipeline and visiting a village close to the site of the explosion that occurred on the Turkish section of the pipeline a few days before the start of the war. In *The Oil Road*, the authors present themselves as eyewitnesses to events, taking the reader as close as possible to the pipeline itself and the hidden processes that their travels bring to the surface (cf. Foster 1995). In this rare and detailed account of the practice of what I have termed political fieldwork, Marriott and Minio-Paluello are consistently critical of the operations of transparency and corporate social responsibility, drawing attention to facts that have only been revealed by whistle-blowers, and understanding the oil corporation's commitment to corporate social responsibility as a type of political management. Recalling a conversation with a Turkish employee of BTC, whose work they nonetheless commend, they observe that her story 'ultimately reveals what her job is intended to achieve: in softening the edges of the twenty-first century Oil Road, [her] team legitimizes the pipeline, monitors and prevents opposition' (ibid.: 221). The corporation's talk of social responsibility and transparency is portrayed ultimately as a form of ideology that the critical researcher must dig beneath (cf. Carroll 2012: 282).

While the authors criticise practices of corporate responsibility and transparency, it is striking the degree to which they valorise the experience of ordinary people as alternative sources of knowledge. This position becomes evident, for example, when they are reluctantly drawn into a discussion with an EBRD specialist about the technical details of a case centred on the impact of the pipeline on a resident of the village of Qarabork in Azerbaijan, whom they had previously met. The resident believed that the pipeline would pass under her house; at the same time she had been told that 'they [BP] won't give us any compensation' (Marriott and Minio-Paluello 2012:

103). In the meeting between Platform and the EBRD specialist, the latter was primarily interested in the technical aspects of the case, including the precise path of the pipeline, the thickness of the pipe walls and any potential risks associated with the technique of 'Horizontal Directional Drilling' (HDD). This was a technique that had been adopted by BTC in order to enable the pipe to pass deep beneath the ground and which therefore, it was claimed, avoided any potential damage to the properties in the village. By contrast, when recalling this scene, Marriott and Minio-Paluello stress the importance of attending to the experience of the villager:

> we should have resisted arguing [the villager's case] on technical grounds, where the masterly self-assurance of numbers renders everything solvable … we should have said [to the EBRD specialist]: 'But if BTC are so sure it is safe, why haven't they properly informed the householders? Why are they so terri-fied? Why does she *want* to be moved?' (ibid.: 106, emphasis in original)

In this rhetorical reflection, it is the reality of the affective experience of the villager that is stressed by the authors, as well as the failure of the pipeline company to inform her properly in accordance with the IFC's resettlement policy (Decker et al. 2003). Indeed, the villager 'seemed powerless even though BTC and SCP effectively passes through her home' (Marriott and Minio-Paluello 2012: 166). However, equally significant is how Marriott and Minio-Paluello do not refer either to the BTC consultants' observations on this specific dispute and its subsequent resolution (BTC/ESAP 2002c: 123, BTC/SRAP 2003c: A-27, B-13–14, BTC/SRAP 2004c; B-9–10), or to the IFIs' response to the NGOs' comments on this case (IFC 2003c: 12).[4] In this instance, the authors are alert to the role of environmental and social specialists in the project; but in privileging the importance of experience, they place the contents of the archive, including the reports of such specialists, firmly outside the frame of their analysis. I return to discuss the case of Qarabork in the next chapter.

The account of Qarabork is but one of a number of examples. Marriott and Minio-Paluello's travels along the BTC pipeline bring them to a series of other locations, in which the impact of BTC has been equally problem-atic. For example, in Haçibayram, Marriott finds that BTC consultants claim to have 'consulted' a village by telephone, when the putative village was actually unoccupied (Marriott and Minio-Paluello 2012: 202; see Chapter 5). In Atskuri, in south-west Georgia, a villager tells them how the vibration from contractors' trucks 'caused regular landslides and rock-falls onto the homes below' the road (ibid.: 173). In Gölvasi, near the western end of the pipeline on the Mediterranean, they learn that the best fishing fields have now become the site of the oil terminal at Ceyhan (ibid.: 236). In detailing this series of cases, the authors consistently present an abduc-tive argument. In themselves, none of the cases are particularly remarkable;

individually, they may not amount to that much. But when they are assembled together, they are cumulatively held to amount to what Peirce termed a 'surprising fact' (Peirce 1934: 117). The surprising fact is that despite the stated commitment of BP to the principles of transparency and corporate social and environmental responsibility, these events actually happened. Collectively, they become indices of wider and deeper problems, directing the reader 'to suspect', to use Peirce's term, that the public commitments of the company are not to be believed or are shallow. In this way, the abductive logic of the authors' account serves to perform its own context. And in this way, given Marriott and Minio-Paluello's understanding and representation of them in *The Oil Road*, the abductive inferences enter performatively into the public debate, contributing to the continuing transformation of a political situation. It is a political situation in which the ethical conduct of oil corporations is brought into question.[5]

Platform's critical explorations of the political geography of the oil industry can be compared with another form of experimental engagement between art, research and corporations. This was an influential approach taken by artists to researching corporations developed in the 1960s by the Artists Placement Group (APG). The work of APG both anticipates and differs from the later practice of Platform. In 1966, APG began to invent a type of art practice that involved working inside selected corporations and institutions including, amongst others, the National Coal Board and ESSO (Corris 1994, Bishop 2012, Tate 2012). Upon entering a particular private corporation or government bureaucracy, the artist associated with APG operated without preconceptions, acting as an employee, consultant or researcher and participating in the everyday life of the organisation. The artist was understood by the group's founder, John Latham, as an 'incidental person' whose presence and actions might effect change. In pursuing these engagements, APG sought neither to represent nor to critique the corporation or bureaucracy from the outside, but 'to introduce change ... through the medium of art relative to those structures with "elected" responsibility for shaping the future – governments, industries and academic institutions' (Barbara Steveni quoted in Walker 2002: 55). For APG, the negotiation of access was itself part of the practice of art. The conceptual art of APG was thus an art of process, of events and of unpredictable effects, a dematerialised art (Lippard 1973), although it was also concerned with objects and materials. Art was to be practised not through opposition or antagonism to the institution at issue, but through immersive participation in organisational life, affecting or reinflecting the organisation from the inside (Bishop 2012: 163–177). In these circumstances, the very fact that bureaucratic or corporate institutions allowed artists to participate in the working life of the organisation without preconditions was an indicator of the success of APG, and was itself already a change.

Some of those who have followed the work of APG have distanced themselves from what they perceive to be the compromised or reformist character of its practice. Already, by the 1970s, APG was being criticised for failing to side with the position of the workers, or for acting as consultants for management (Walker 2002). At a conference organised to mark the acquisition of the APG archive by Tate Britain in 2005, a younger generation of artists, including a member of Platform, spoke about their own attempts to engage with the relation between art and corporate and bureaucratic institutions. While they acknowledged the influence of APG, this new generation preferred either to use art to criticise institutions from the outside, or to act critically from the inside in clandestine ways. They explicitly rejected the idea that artists should openly negotiate access to institutions over a long period and adopt the role of researchers or consultants. One element of Platform's guided tour of the oil economy of London took place in a rented city office, in this way performing the act of going inside the corporation while retaining the perspective of the external critic.

Where Platform have taken up a position external to the corporation, engaging in a form of critique, APG explicitly resisted the language of radical politics and refused to adopt a 'Frankfurt School orthodoxy of apartheid between artists and government' (Latham 1986: 49, quoted in Slater 1999). Indeed, APG sought to effect change through means – inhabitation, consultancy, long-term research – that are not conventionally viewed as political. The position of the APG artist was close to the position of the bureaucrat, manager or consultant – close enough that one commentator suggested 'there is, in the organizational "unconscious" of the APG, a mindset that seeks legitimation for an art practice not from the art institutions themselves but from industrial and government professionals' (Slater 1999: 3). The difference that an artist can make would be achieved through proximity to government or the corporation, not through opposition; it would be unpredictable, and possibly infinitesimal. The result would be an inflection in the curve of the evolution of the institution, not a rupture with the past (Tarde 2001 [1890], Deleuze 1993: 47).

The practices pursued by Platform and APG are therefore markedly different, or even opposed. Adopting an explicitly ethical and political agenda, Platform carries out research externally and in opposition to the corporation: in this case, BP. In contrast to the practice propounded by APG, in which difference was to be fostered through a sustained immersion in a corporate milieu without a pre-ordained politics, in part through collaboration with corporate professionals, Platform's approach entails the primacy of a mode of oppositional politics. The contrast between APG and Platform mirrors the contrast between two ways of thinking about the relation between business and ethics that I introduced earlier in this chapter. On the one hand, artists associated with APG worked alongside and with the professional employees of large public and private corporations. In effect,

APG artists sought to foster change within the types of bureaucratic and professional organisation that became established over the course of the twentieth century – organisations that can be expected to have contained the kind of professional ethos that was identified by Durkheim. In contrast, Platform's practice challenges the corporation from the outside, reflecting and responding to the now-dominant paradigm whereby the ethical conduct of corporations has publicly to be demonstrated to others in accordance with the requirements of international law, standards and guidelines (Thompson 2012). For Platform, art and research become subordinate to a given politics; whereas for APG, art has an ethic that is irreducible to politics. Indeed for APG, we might say, 'the creativity of what is new and the production of autonomy take on an exemplary importance precisely in so far as those are the problems that intrinsically confront all artists in so far as they are artists as opposed to anything else' (Osborne 1998: 123).

Conclusions: Ethics and Knowledge Controversies

The ethicalisation of the oil industry during the 1990s and 2000s took the form, as we have seen, of the development of a growing body of expertise in fields including reputation management, human rights monitoring, environmental auditing, and environmental and social impact assessment. This entailed the production of increasing quantities of information in order both to meet the demands of investors and civil society organisations and to address a growing range of regulatory requirements and guidelines. The BTC pipeline was promoted as a test case or a model for this emerging form of ethical capitalism. However, these changes cannot be reduced to rule by 'enlightened specialists'.

In this chapter I have contrasted two approaches to the generation of evidence about the ethical conduct of the oil industry. On the one hand, the industry itself employs a burgeoning cadre of consultants to audit, monitor and assess its ethical performance. As we shall see, the construction of the BTC pipeline entailed the development of a particularly elaborate apparatus of institutionalised reflexivity (Born 2005a). On the other hand, radical critics of the oil industry, rather than simply presenting a general critique of its ethical conduct, have engaged in political fieldwork in order to document specific instances of unethical conduct, so rendering them particularly visible. Demonstrations of the ethical conduct of corporations are challenged by counter-demonstrations. The routinised production of information by the industry is challenged through the use of examples along with an attendant abductive logic. In these circumstances, disputes about ethics (are corporations acting ethically or not?) frequently take on the form of disputes about evidence (doesn't this evidence show that they are not acting ethically?). In short, the ethicalisation of the oil industry creates particularly conducive grounds on which knowledge controversies are fomented and can flourish.

In the introduction to this book I argued that it is productive to supplement the analysis of knowledge controversies with a new concept, the political situation. The notion of the political situation is intended to convey a sense of the indeterminacy of the boundaries and limits of particular knowledge controversies. Focusing on the work of Platform, we can now see two good reasons for adopting this approach. First, specific disputes, such as the modest disputes that revolved around the villages of Qarabork, Atskuri and Haçibayram, may not have had great political significance in themselves. However, such events can be made to carry much greater significance when they are framed as mere elements in a larger constellation of events. When placed alongside a number of apparently similar controversies, the conflicts over drilling in Qarabork or traffic movements in Atskuri accrue a new vitality, becoming elements of a surprising fact. The surprising fact is that the corporation routinely fails to act in the way it claims to act. It does not properly inform the resident of Qarabork of the risks (or lack of risks) of drilling under her property, or take adequate care about the potential impact of pipeline-related traffic on Atskuri. In short, BTC does not conduct itself responsibly: it espouses the principles of corporate social responsibility, but it fails to enact them. Thanks to the work of Platform, it is no longer the specific knowledge controversy that is at stake, but the larger series of controversies within which the individual controversy is nested or placed.

The second reason for adopting the concept of the political situation follows directly on. It is that the political context within which particular cases such as Qarabork or Atskuri are framed is itself also not given. It is unclear whether these controversies are just about specific problems such as the impact of drilling or traffic, or whether they can be taken to be indices of more profound or wider problems: that is to say, as elements in a political situation, it is precisely their scale, scope and boundaries that are at stake. Indeed, while the IFIs and the BTC company recognised that a dispute had occurred in Qarabork, they did not acknowledge that it had any wider implications (Chapter 5). In contrast, Platform's abductive logic directs its audience towards the importance of a wider political controversy – about the ethical conduct of corporations in general – to which the cases of Qarabork and Atskuri make a contribution. By enacting or performing this framing, the space of controversies is enlarged and extended, becoming delocalised. But in addition, through the logic of abduction, the political situation within which the significance of the particular controversy is held to make sense may itself be transformed. The logic of abduction is, then, a central mechanism in the unfolding and in the transformation of political situations. It points at once to the perspectival, relational nature of political situations, to how particular knowledge controversies are both understood to be and enacted as enmeshed in them, and to how the very enactment of such an understanding is performative: how it then contributes to the further transformation of both particular knowledge controversy and political situation.

In the next chapter, I turn to consider how the ethicalisation of the oil business directed the attention of investors, auditors, consultants and critics not just to the political importance of particular materials and locations, but to the generation of social forms: interested publics and what were called 'affected communities'. If material entities such as pipes and traffic played a critical part in the ethicalisation and politicisation of the pipeline, the significance of such entities was entangled with the constitution of communities that they were thought to affect. In these circumstances, civil society, stakeholders, affected communities and the public were all thought to be able to make an important contribution to the development and design of the BTC project. As we shall see, however, the challenges entailed in constituting a series of stakeholders in relation to a physical infrastructure, and particularly one that is 1760 km long, are far from obvious. To understand the proliferation of disputes around the pipeline, we need to attend to the ways in which the corporation sought to manage the impact of its operations by determining the limits of its responsibility to interested publics.

Chapter Five
The Affected Public

The construction of the Baku-Tbilisi-Ceyhan (BTC) pipeline involved much more than engineering, finance and the capacity to negotiate agreements with national governments. It was expected that the social and environmental impact of the pipeline would be assessed and mitigated, and that affected populations would be both informed and consulted. In particular, the project would be governed by the requirements of the World Bank Operational Directive on Environment Assessment (World Bank 1999), the European Bank for Reconstruction and Development (EBRD)'s policy on public information, the European Commission directive on 'Freedom of Access to Information on the Environment' (EEC 1990) and the Åarhus convention 'On Access to Information, Public Participation in Decision-Making and Access to Justice in Environmental Matters' (UNECE 1998). The latter stipulated, for example, that affected populations and other interested parties should be treated as a 'concerned public' (ibid., article 2) which should have an opportunity to participate in decision-making (BTC/ESIA 2003: 13–17).[1]

These were potentially challenging demands, not least because of the scale of the project, the political instability and weakness or absence of democratic political institutions in the region, and the limitations of relevant social scientific research on eastern Turkey, Georgia and Azerbaijan (Grant and Yalçin-Heckmann 2007: 5). They represented an additional challenge because these evolving policies had yet to be fully tested on a project of this scale and complexity. While the International Finance

Material Politics: Disputes Along the Pipeline, First Edition. Andrew Barry.
© 2013 John Wiley & Sons, Ltd. Published 2013 by John Wiley & Sons, Ltd.

Corporation was able to build on its recent experience of the Chad-Cameroon pipeline project (IFC 2003b, 2006), the European Bank for Reconstruction and Development had not, at this time, had substantial experience of public engagement. At this time, as one informant noted, EBRD had an (economic) 'transition mandate' but not a 'social mandate'. Likewise, BP itself had little in-house expertise in social research – as distinct from economic and political analysis.[2] In effect, for the oil companies and for the international financial institutions (IFIs), the construction of the pipeline, a structure that crossed three countries and was financed through a global network of financial institutions, posed a series of problems. How was it possible to identify, let alone inform, consult and engage with a public or publics stretched along the length of a 1760 km corridor of land? What would they be informed and consulted about? How, and to what purpose? Apart from those immediately affected by the construction of the pipeline, who else might act as spokespersons for these publics? In what ways could the existence of such publics and their concerns be verified and eventually made public? In this chapter, I trace the multiple ways through which BTC and the international financial institutions sought both to assemble and to represent such publics, which were defined not by their citizenship of a state, but through their possible relation to an emerging object, the pipeline. In particular, at the heart of the BTC project was the biopolitical problem of how to consult and respond to the concerns of a specific public. This was the population living in the vicinity of the pipeline, who came to be known as 'affected communities' (cf. Petryna 2002). In turn, the oil company and the IFI's claims to have properly assembled and addressed the pipeline's publics were disputed.

These problems, and the dilemmas facing the consortium responsible for the project, point us towards older ideas about publics, how they should be conceived, and how they can be assembled. In his essay 'Ideology and Ideological State Apparatuses', Louis Althusser proposed the notion that the relation between the ideological function of the state and the individual subject could be understood as a form of interpellation: ideology served to interpellate or hail the individual, and to constitute the individual as a subject (Althusser 1984, Warner 2002: 67). However limited, Althusser's account provides a useful starting-point for thinking about the problem of how to assemble the public. The public has in part what we might call a governmental existence (Foucault 2007, Burchell et al. 1991, Barnett 2003: 83–84). In all its diverse forms, the public is continually hailed or addressed, whether through the enunciative acts of politicians (as 'the American people', for example) (Latour 2005b); by institutions that claim both to reflect and to constitute national and regional publics, for example, public service broadcasting (Donald 1992, Born 2005a); or through the routine use of devices such as public opinion polls (Osborne and Rose 1999) and elections (Barry 2002, Coles 2004, Johnston and Pattie 2006). One could say that the public is expected to be both interested in politics and to take an interest in its own political existence. At the same time, the public in a liberal democratic society should not be forced into existence or subjected to propaganda. The public is addressed and named, but it is also expected to be sufficiently affected that it freely participates in this process, that it voices its own opinion, that it accepts or contests the claims of political parties that they can represent the public, and that it recognises or challenges the address of national (or local) institutions.

The idea, or ideal, of the public is therefore characterised by a set of expectations that may be given more conservative or radical formulations. The expectations placed on the public can be limited or extensive, infrequent or continuous, demanding or relaxed. Nonetheless the public is understood as an entity, which *should* make itself manifest in a democratic society (e.g. Dewey 1927, Dryzek 2000). Publics have often, of course, been equated with national publics, and they are addressed by, or address themselves to, national political institutions. But publics are also increasingly called upon to address, and themselves mobilise in relation to, problems, issues or objects that transcend national or regional boundaries (Barry 2001, Beck 1999, Marres 2005, Marres and Rogers 2008, Jessop 2008, Laurent 2011). In these circumstances, the problem of how to assemble a public, which can no longer be assumed to be contained within a given national space, region or territory, frequently has to be confronted and solved afresh (Amin 2004).

It is common to assume that the public has an immanent existence, waiting to be addressed and activated, only constrained by the absence of appropriate liberal democratic safeguards, or by its self-disciplining sense that it is only proper to act as a public when called upon to do so. There is a national public, in this view, and its unity and existence can be taken for granted, only to be consulted, more or less imperfectly, on appropriate occasions (Warner 2002: 65). But if we consider publics as collectives that are called into existence in multiple forms and spaces, then our attention is necessarily drawn to the diverse techniques employed both to assemble and to speak on behalf of specific publics (Latour and Weibel 2005). Some of these practices and settings have a continuing existence over time and may be institutionalised. They are associated with particular forms of speech, employ specialist forms of expertise and technical devices, and may involve well-developed procedures as to how they should be used. Parliamentary democracy, for example, has its own technology, not only in the guise of the apparatus of elections (ballot papers and machines, polling booths) but how votes are cast and counted in parliament and the extent to which representatives are constrained or not by the directives of their party leaders and managers. There is an inevitable arbitrariness and path-dependency to the history of such procedures, settings and technologies (Dányi 2011). But such arbitrariness is not a disadvantage as long as those involved accept that such procedures enable democratic systems to reach decisions on matters on which there will inevitably be disagreement (Waldron 1999a). If parliamentary institutions rely on the legitimacy of tradition, however arbitrary and country-specific, other ways of assembling publics may also need to be invented, improvised or copied from and elaborated on the basis of practices that have been tried elsewhere.

I want to suggest here that it is productive to suppose the existence of generic forms of public-making, that is, ways of assembling publics and of gauging and articulating their will or opinion. In making this suggestion I draw an analogy with genre theory in literature and film. Literary and film theorists have long recognised that the notion of genre does not just apply to the study of texts in themselves (Neale 1980, Frow 2006). Certainly, texts invariably participate in genres that exist and mutate over time (Derrida 1980: 230). Film historians, in particular, have focused not solely on the work of individual directors, but on generic forms: the action film, the western, the romantic comedy, the social realist drama and so on. But the existence and multiplicity of such generic forms also points to the existence of producers and audiences who understand the generic conventions and who both follow and influence generic transformations over time. Genres, in other words, are critical to understanding the shifting 'orientations, expectations and conventions' between texts, audiences and producers (Neale 1980: 19). Genres condense bodies of convention, but the conventions are not static or given and are themselves subject to interference, spatial variation and processes of change.

Today, the array of 'genres' of public-making is quite extensive and diverse: it includes the town hall meeting, the TV debate, the public inquiry (Ashenden 2004), the opinion poll (Osborne and Rose 1999), the organised campaign (Sadler 2004), the stakeholder forum, the public consultation process (Lezaun and Soneryd 2007), the participatory technique (Callon et al. 2001, Davies 2006, Cooke and Kothari 2001, Whatmore 2009, Laurent 2011, Marres 2012), the transnational network (Riles 2001, Andolina et al. 2009), the occupation, the strike and the march (Tilly 1986, Barry 2001, Amin and Thrift 2005, Featherstone 2008, Mitchell 2011). All of these forms and techniques, which have widely varying political significance and visibility, are understood as ways of assembling and performing publics, whether through the presence of representatives or through the representative presence of members of the public. The analogy between politics and film and literature draws attention both to the historicity of particular genres, and to their re-invention and mutation in different circumstances and settings. It directs us towards the existence of a great diversity of ways in which publics are assembled and speak or are spoken for, and the need to identify and interrogate these specific means in relation to any genre. It points also to the need to consider the extent to which particular forms may be experienced by actors as more or less participatory or egalitarian – that is, exclusive or inclusive — in practice.

In this chapter, I focus on three generic practices of public-making in turn, all of which were deployed during the development of the BTC pipeline, across a range of settings. They are: 1) the public disclosure of information, 2) procedures of environmental and social impact assessment, and 3) the stakeholder forum. Each of these forms involves practices that should be understood as performative, and they were certainly intended to be so (cf. Muniesa et al. 2007). None of them addressed pre-existing collectivities. Rather, they were expected to assemble and address new collectivities – 'civil society', 'affected communities' and 'stakeholders' – that should have an interest in the construction of the pipeline. These collectivities were understood to be social groups defined by their relation to an evolving object, not by reference to their membership of a state (Harvey 2010, cf. Jeffrey 2013). While I address all three in this chapter in turn, I focus in particular on how the BTC company and the IFIs sought to assemble and address affected communities, conceptualising them in terms of their location within a narrow corridor of land on which the pipeline construction was expected to have an 'impact'. Mapping the corridor of impact provided a solution to the biopolitical question of which communities would count as 'affected'. At the same time, the effort to determine what the impact on affected communities would be itself had an impact. The designation of a village as an affected community could have a second-order effect (Luhmann 2002, Esposito 2011), for a community was likely to be affected by the observation that it might be affected in the future.[3]

Yet if these generic forms of public-making were expected to generate empirical knowledge about publics, their opinions and concerns, they also became the focus for criticism. Some of these criticisms relied on the generation of empirical political knowledge, gathered through 'fact-finding missions' by international NGOs, who acted as counter-experts, challenging the company's account of who was affected and what were the actual concerns and interests of affected communities (Blok 2007). But as sociologists of scientific knowledge lead us to expect, the critics also questioned the competence and trustworthiness of those experts employed by the corporation and the IFIs who claimed, on the basis of research, to represent the affected communities' problems and concerns. The critics therefore challenged whether knowledge of affected communities had been properly generated in the first place. In this way, a public knowledge controversy developed around the question of the constitution of affected communities as an object of knowledge, and whether oil companies or international NGOs were best placed to act as spokespersons on their behalf (cf. Latour 2005b: 31). Such disputes about the constitution of the pipeline's publics were themselves played out in public. But an exclusive focus on what was made public in these disputes would provide a misleading view of the dynamics of the controversy about the constitution of these publics. While disputes might appear to be controversies about matters of fact, which could in principle be resolved through further research and the generation of further facts, in practice many informants viewed these disputes as manifestations of what was fundamentally a series of antagonistic relations between the oil companies, the IFIs and their critics. In this context, there was no possibility of a resolution, whatever further facts might have been generated by either side. Disputes about matters of fact had to be understood in the context of underlying problems and conflicts – ongoing political situations – that transcended any particular issue. Thus, for many of those involved, the very significance of the disputes was undetermined.

Disclosure

During the late summer of 2003, a young Czech environmentalist, Martin Skalsky, made a tour of Azerbaijan, visiting the offices of the EBRD in Baku, the Baku Enterprise Center and the District Executive Authorities Office (DEAO) in Yevlakh in the west of the country. Each of these institutions was designated as a centre at which documentation concerning the 1760 km oil pipeline that was planned from Baku on the Caspian Sea to the Turkish Mediterranean coast, should be available to the public. Similar documentation was available to be consulted at the offices of the EBRD on Liverpool Street in London, as well as various locations in Turkey and Georgia. In order to receive financial support from the International Finance Corporation (IFC), the BTC company was expected to comply with IFC policies on information disclosure and consultation policies, as well as with the EBRD's policy on consultation, European Commission directives, national (i.e. Turkish, Azeri and Georgian) regulations and rights established by the Åarhus convention (World Bank 2000, IFC 2003a, Baku-Ceyhan Campaign 2003a&b, Barker 2006: 115, Jeter 2006).[4] Before the decision on financing by the IFIs was confirmed in late 2003, a final 120-day period of public information disclosure and consultation was required. Forty-six volumes of documentation, totalling 11,000 pages, were made public at this time (IFC 2003a: 19, Pollett and Wyness 2006).

On his tour, Skalsky took notes about the availability of documentation on the pipeline in these offices, whether it was in Russian or Azeri, the helpfulness of office staff, and whether they were knowledgeable enough about the documents to answer visitors' questions. He noted that the Baku Enterprise Center had a pleasant and comfortable environment and had an accessible computer with a data projector. At the EBRD offices there was the possibility of free copies, and tea and coffee for visitors. However, in the Enterprise Center, at one of his first meetings with local non-governmental organisations interested in the development of the pipeline, participants said that they were watched either by BTC company or state security. In Yevlakh the situation was worse. Police guarded the entrance to the DEAO and every visitor had to be registered. A DEAO representative claimed that they had consulted 3000–5000 citizens of Yevlakh and neighbouring districts and that 200–300 citizens had come to familiarise themselves with the documentation. The representative claimed that 'everybody' was satisfied with the process. Given that he estimated that no more than 100 people had looked at the documentation in Baku, and Yevlakh was a much smaller provincial city, Skalsky reckoned that the official's claims were implausible: 'it is very strange that the local co-worker of the EBRD and the International Finance Corporation (IFC) has clearly lied about this ... [the] case backs the suspicion that the documentation in Yevlakh was not accessible at all' (CEE Bankwatch 2003: 9). It was strange too, he thought, that the documentation was placed in Yevlakh, and not in Ganja, the second largest city in Azerbaijan that, unlike Yevlakh, possessed a university, an Academy of Science, and one or two NGO resource centres. At least in Ganja, he claimed, it was more likely the documents would have been read.

Skalsky's research project is instructive because he was not at all concerned, at least in the report that he produced, with the content of the copious documents. He was preoccupied, rather, with the material conditions under which they could be accessed, and whether they had been placed in a setting where members of the public could become informed – as the company and IFIs had promised. He was interested, in other words, in the performance of a specific generic form: the technique of public information disclosure and its use in the field (CEE Bankwatch 2003). He did not seek to judge the value of this genre of making things public in general, but rather criticised a particular example on the basis of his own observation in the field. He was engaged in the generation of empirical political knowledge, on the basis of which he could raise, but not answer, the question of why documents were placed in Yevlakh rather than Ganja.

However, Skalsky had only observed one specific moment of a much more complex process. For over the course of the previous two years, BTC had contracted a vast enterprise of social and environmental research and public consultation along the route of the prospective pipeline, in order to comply with, or even to go beyond, the demands of international and national guidelines and regulations. This led to the preparation of a Resettlement Action Plan (RAP), an Environmental and Social Impact Assessment (ESIA), a Public Consultation and Disclosure Plan (PCDP), as well as further reports dealing with a series of technical and environmental issues, along with documents relating to the construction of the South Caucasus Pipeline and the Azeri-Chirag-Guneshli oil field. It was these documents that formed an important fraction of the vast quantity of material available for inspection in Baku and Yevlakh.

Over the course of the previous two years, public consultation and disclosure was carried out not just in cities such as Baku but in a range of locations, at a range of scales – local, national, international – forming an element of complex assemblage of multiscalar governance (Jessop 2008, Chapter 9). Moreover, the process of public disclosure and consultation addressed a whole series of audiences including *inter alia* government ministries and regulators, regional and district authorities, national NGOs and donor organisations in the region, regional offices of international organisations and NGOs, universities, academies of science, conservation groups, charities and international NGOs, such as BirdLife International, Friends of the Earth, and the World Wildlife Fund for Nature (e.g. BTC/ESIA 2002 a, b, c & d, 2003 a&b, BTC/RAP 2003). But at the heart of the project of public consultation and disclosure were the communities that were likely to be affected by pipeline construction and operation (BTC/ESIA 2002e). It was expected that the communities themselves would be consulted, and disputes along the route of the pipeline revolved, in part, around the question of who had the capacity and the competence to speak on their behalf.

Affected Communities

Affected communities did not pre-date the development of the pipeline, but villages lying near to the route of the pipeline became affected by the project long before construction work began. Indeed, there was a strong sense that the idea of an affected community was performative; for the company had to bring affected communities into being in order that they could be informed and consulted, and the impacts on them assessed. A footnote in the final version of the BTC Environmental and Social Impact Assessment report for Georgia provides us with the following brief definition of what was considered to be an 'affected community':

> Pipeline affected communities are defined as those that are located within (or partly encroach into) a 2 km corridor either side of the route, or are within 5 km of a potential worker camp or pipe yard. These communities are likely to experience and be affected by the activities of construction, operation and decommissioning of the pipeline. (BTC/ESIA 2002a: 1-35)

In this definition, the identity of an 'affected community' is determined by its distance from the future route of the pipeline itself, as well as worker's camps and pipe yards, the presence of which was expected to be particularly disruptive for those living in their vicinity.

Yet this identity had to be forged progressively. As early as January 2001, consultants began to visit all 69 villages in Georgia, speaking with village leaders, and interviewing 620 households, even prior to the main phase of

community consultation and public disclosure of the draft ESIA (BTC/
ESIA 2003: II-22). There were repeated further visits. In this way, whatever
the skills of the environmental scientists, geologists, archaeologists, anthropologists and sociologists engaged in the conduct of environmental and
social impact assessment, the assessment itself had an impact. Not only
were local populations informed that they were part of an international
project to bring oil from the Caspian Sea, fostering expectations of potential
gain (Ferguson 1999, Weszkalnys 2008), they were also made to appraise
the construction of the pipeline in terms of its 'environmental and social
impact', and how this impact, and their existence as affected communities,
might be mitigated and compensated.[5]

However, the idea that it was only the construction of the pipeline itself
that impinged on the lives of the affected communities proved difficult to
sustain. The pipeline corridor was not a closed social system. In Georgia,
informants observed that the Schevardnadze government had fostered high
expectations of the benefits of the project for the country, including the
generation of thousands of jobs along the route of the pipeline, which would
never materialise.[6] Yet whatever those engaged in the conduct of environmental and social impact assessment learned through the course of their
research, the ESIA and RAP contained little analysis of the activities of the
Georgian, Azerbaijani and Turkish state or state oil companies in the villages
along the pipeline route. Moreover, while the ESIA reported social and
environmental impacts exhaustively, it provided little account of the lived
experience of political and economic life in post-socialist Georgia or
Azerbaijan, or what Humphrey has termed the 'unmaking' of socialism
(Humphrey 2002, Yalçin-Heckmann 2010). In this light, the few observations that are made in the BTC archive about the relations between state
and society are striking in both their rarity and their discretion. One BTC
report, for example, noted that: 'the Government of Azerbaijan has moved
to strengthen district and local government with the first local elections to
establish municipal administrative bodies held in 1999. In practice, the
roles and responsibilities of the municipalities and their relation to district
administrations are still being defined. Lack of resources has lead to widespread reliance of both the public and private sector and private citizens on
informal networks and systems of payment' (BTC/RAP 2002d: 4–5).

The report can be contrasted with the account given by an international
NGO, which alleged that local authority executive committees (in
Azerbaijan) were appointed by the state and represented and provided
effective means of exercising government control throughout Azerbaijan
(International Alert 2004: 51). At the same time, the Georgian NGO Green
Alternative argued that 'IFI support for [the BTC project] does not facili-
tate the alleviation of poverty and corruption' (Green Alternative 2005).
By constituting 'affected communities' the company forged a distinction
between, on the one hand, a society on which the pipeline impacted and, on

the other hand, the complexity of the relations between such communities and the state (Yalçin-Heckmann 2010), which was not addressed in published documents. Even when BTC published a wider review of the politics and economy of the region, the question of the relation between the activities of local and national government, on the one hand, and the impact of the pipeline construction on affected communities, on the other, was addressed only briefly (BTC/RR 2003).

Thus the affected communities were to form a distinct social and political order that was contained within, but conceived as purified from, the political order of the state. Provision of information to affected communities should produce 'a high level of awareness among [such] communities and other stakeholders about the nature of the project, its likely impact and proposed mitigation measures' (BTC/ESIA 2003: II-18)[7]; while, at the same time, the company aimed to contain its dialogue with the local population within strict social geographical limits, thereby 'avoiding raising expectations in a large number of dispersed communities' (BTC/ESIA 2002c: 9-1). Through the provision of information and through extensive consultation, expectations could be both managed and limited, and problems addressed. In this way, the formation of affected communities was critical to the attempt to constitute the pipeline route as a space of transnational government that was distinct from the territory through which it passed (Foucault 2007). I return to consider the contestability of the distinction between affected communities and the state in Chapter 8.

The brief definition of 'pipeline affected communities' given earlier conceives of the identity of such communities simply in terms of their distance from the route of the pipeline. But in practice the corridor of pipeline affected communities formed part of, and was internally divided into, a series of corridors, the existence of which would affect communities in distinct ways.[8] These corridors included: (i) a 10 km wide 'corridor of interest', which had been identified in a 'multi-disciplinary study' in 2000; (ii) a 500 m wide 'preferred corridor', within the corridor of interest, which was submitted to the governments of Georgia, Azerbaijan and Turkey in 2001; (iii) a 100 m wide 'specified corridor', based upon preliminary environmental, social and technical studies; and (iv) a 44 m wide 'construction corridor' which would be the starting point for detailed assessment by social specialists (BTC/ESIA 2002b: 4-4). In turn, the construction corridor would be leased during construction (in Azerbaijan) or purchased (44 m wide in Georgia) or partly purchased (8 m wide) and partly leased (28 m wide in Turkey). In addition, these corridors co-existed with a series of other corridors governing social and economic matters, including public information disclosure and stakeholder consultation. These included: (v) a corridor 2 km either side of the pipeline (or 5 km from a pumping station or major AGI) within which a socio-economic survey was conducted, and which formed part of the ESIA (BTC/ESIA 2002b: 7-2), thereby creating approximately one hundred affected communities, which was considered a manageable number; (vi) a 4 km wide corridor within which the Community Investment Programme (CIP) would support projects intended to foster the long-term social, economic and environmental sustainability of the community (BTC/ESIA 2003); (vii) a 'pipeline protection zone' (58 m wide in Georgia and Azerbaijan) on which there would be 'restrictions on use' including some kinds of tree planting, ploughing deeper than 30 cm, constructing animal pens, and the use of explosives (BTC/RAP 2002d: 2-3); (viii) a Right of Way [ROW] within which regular horseback inspections would be carried out to ensure that the length of the corridor remains 'tree-free' throughout the pipeline's operational life (BTC/ESIA 2002d: 11-30); (ix) a corridor 15 m each side of the pipeline within which habitable buildings were prohibited but allowed for normal agricultural use (BTC/ESIA 2002c: 5-4); (x) a 500 m 'broader zone' either side of the pipeline within which 'major developments' such as schools and hospitals would be restricted (ibid.); and (xi) specific corridors governing compensation for losses in production associated with specific forms of agriculture. In addition, the pipeline itself was 42 inches in diameter in Azerbaijan and Turkey and 46 inches in Georgia (BTC/RAP 2002d: 2-3).[9]

These different corridors were the objects of different forms of expert analysis, by construction engineers, specialists in land acquisition, social scientists, environmental scientists and even archaeologists: 'an archaeologist with a watching brief will accompany the construction activities on the pipeline, and will record the presence of archaeological features ... [and] the appropriate response will be decided upon in consultation with the Ministry of Culture and Institute of Archaeology and Ethnography [of Azerbaijan]' (BTC/ESIA 2002b: 10-45). The corridors were linked to a vast range of specific 'commitments' on the part of the company including a commitment – frequently mentioned by informants during fieldwork – to restore the pipeline corridor to something close to its original state. Different corridors had different forms of public visibility and were subject to different forms of scrutiny, as well as to different processes of consultation with affected communities and other stakeholders. Whereas, with exceptions, little would have been known, except by construction workers, engineers and managers, about the engineering work that would take place within the construction corridor, many came to know what price the oil company would be prepared to pay in compensation for losses in agricultural production, for example. Nonetheless, as we shall see in later chapters, the rationale for the widths of particular corridors was not necessarily clear, not least to the residents of affected communities themselves. It is perhaps not surprising that although they were apparently distinct, these corridors could become conflated and the question of the relation and distinction between different corridors created effects, generating further questions. Why would a village be consulted but not compensated, for example? Did the level of compensation reflect the risks that the pipeline posed to the health of the local population? And what was the relation between these corridors – the width of which was defined with such precision – and the impacts that would affect communities in practice? I return to consider the disputes surrounding the question of what constituted a real 'impact' in Chapter 6.

The corridors themselves divided up affected communities in a variety of ways. They established distinctions, in principle, between those who had the right to speak as subjects of consultation, those that might be eligible for the receipt of community investment, those who could suffer particular impacts such as traffic noise, and those that owned or made use of land along the route. Moreover, these were not fixed corridors. Formally, they all only existed for periods of time – during the process of planning, during construction, or for the 'lifetime of the BTC project' (BTC/ESIA 2002k: 7-2). The 'preferred corridor' only existed for a brief period, although its existence would have affected those who knew of its path prior to the publication of the ESIA.

Moreover, in certain areas these corridors had to be moved: 'as sections of the pipeline have, from time to time, been rerouted in part as a response to ESIA results, the "surveyed communities" no longer correspond 100% to the "pipeline affected communities"' (BTC/ESIA 2002k, 7-8). In this way, the expectation that the pipeline would bring wealth to affected communities could be subsequently dashed when communities discovered that they were no longer considered as 'affected'.

> ... at both Garabork and Chiyni [in Azerbaijan] there is no doubt that the households with the land in the 'pinch point' areas had high expectations with respect to compensation, and that there was considerable disappointment when it was decided to utilize HDD [horizontal directional drilling] thus not affecting existing land use. This was exacerbated when compensation payments were finalized with nearby households with land in the pipeline r-o-w [right of way] beyond the planned HDD sections. (IFC 2003a: 12)

Indeed, as we saw in the last chapter, Platform was critical of BTC in this case, not just because of their concerns about the potential impact of pipeline construction on the houses of Garabork, but because of the affective impact on their residents who, in Platform's view, could not have been properly informed about HDD and, as a result, were 'terrified' (Marriott and Minio-Paluello 2012: 104, Chapter 4). Ironically, BTC had decided to use HDD in order to avoid the 'resettlement of dwellings' (BTC/ESAP 2002c: 123), in line with the company's wider policy (BTC/RAP 2003: 9). After all, the company did not wish to be seen to be forcing people to leave their homes and HDD appeared to provide a way of ensuring that the residents would not be directly affected by the construction of the pipeline at all. But this technical solution was only a partial success, and itself generated controversy. For the residents had clearly been affected both by the sense that they might be affected by HDD and by the suggestion that they might *not* be considered affected and therefore would not receive compensation. In consequence they were compensated anyway (BTC/SRAP 2003c: B-14), while HDD work was also subsequently carried out (BTC/IEC 2004). In effect, the formation of 'pipeline affected communities' led to the formation of the pipeline corridors not just as a new space of government but as an 'affective space' (Navaro-Yashin 2012).

During the process of public consultation and environmental and social impact assessment research the company did not yet own any of the land. Nor were the pipeline corridors to become an enclave, in the sense of ever being visibly enclosed or demarcated from the surrounding land. The pipeline construction yards, workers' camps, and pumping stations were surrounded by fences, but the pipeline corridors were not (cf. Ferguson 2005). Yet, through the process of public consultation and environmental and social impact assessment, the company and the IFIs progressively established a series of governable spaces; that is, spaces constituted not through physical enclosure but through the production and circulation of knowledge about the pipeline and its impact on, amongst other things, affected communities. In this way, the identity of affected communities was defined not just by their relation to a construction process and a material artefact, but to a series of accounts of these relations, and how they should be managed and rendered visible (Petryna 2002). Affected communities were not necessarily communities, in so far as they experienced a sense of their own unity and identity as affected communities. Indeed, the constitution of the pipeline corridors created internal divisions within affected communities between those landowners who received compensation for the purchase or lease of land and those who did not. Rather, affected communities were multiplicities, enacted as affected communities through their multiple and shifting relations to the construction and operation of the pipeline (cf. Law and Urry 2005).

Points and Corridors

Company documents conceived of the route of the pipeline as a series of corridors – narrow strips of land – the borders of which were defined by their distance from the route of the pipe itself. But as Madeleine Reeves has argued, borders are generally encountered only at points and specific times, not as continuous lines: 'maps suggest a spatial contiguity to sites that may be encountered in ruptures; they suggest a temporal contiguity which may be experienced as sporadic (as in the comment, "we only have a border market today") and they suggest a basic homology to the things mapped (roads, mountains, administrative districts) that are in fact of radically different orders' (Reeves 2008: 78). Reeves' analysis, which derives from her fieldwork in Kyrgyzstan, is concerned with the everyday experience of borders and territories, but her remarks also apply to the experience of the pipeline corridor, and its multiple informational borders. Those engaged in its construction were uniquely able to experience the route as a line by literally driving down it; it became, after all, a right of way for the BTC company and its contractors. By contrast, those who were affected by it encountered it only at particular points or in short sections. In Henri Lefebvre's terms there was a clear difference between the 'conceived space' of the pipeline corridors and the 'lived space' of those who dwelt in its vicinity (Lefebvre 1984).

In 2003 both Amnesty International and the Baku-Ceyhan Campaign had criticised the legal regime governing the pipeline corridors as a whole, as we have seen (Amnesty International 2003, Baku-Ceyhan Campaign 2003a, b & c, Moser 2003). In addition, the coalition of international NGOs that formed the Baku-Ceyhan Campaign directed their attention to the ways in which the public had been assembled and addressed at specific points along the pipeline. Here I focus on the case of the Turkish village of Haçibayram, which assumed a remarkable importance in the dispute between the NGOs, the IFIs and the company over the question of consultation of affected communities. The NGOs' case was based on two key pieces of empirical evidence about the village that had been generated through their own brief 'fact-finding missions' to Turkey. One was that the village, although it was said to have been consulted according to the Environmental Impact Assessment (EIA),[10] was actually deserted when visited by an NGO fact-finding mission: 'the August 2002 FFM [fact-finding mission] found the village, listed in the EIA as consulted by telephone, to be uninhabited, with neither telephones nor residents to answer them' (Baku-Ceyhan Campaign 2003b: 15, see also Baku-Ceyhan Campaign 2003c: 52, Marriott 2007, Marriott and Minio-Paluello 2012: 202). The second was the evidence of a specific witness, the Muhtar [elected head] of Haçibayram, whose evidence refuted the company's and the IFI's claims that consultation had been properly carried out. According to the NGOs he had only once met representatives from the national pipeline company (Botaş) and he had not been contacted by telephone, 'nor had anyone else in his community' (Baku-Ceyhan Campaign 2003b: 15).

The NGOs argued that the conduct of the EIA in Haçibayram was symptomatic of a failure of the company to fulfil the World Bank guidelines on public consultation, and therefore to recognise the concerns of those who might be affected. The case was 'illustrative of the unreliability of the consultation data presented in the EIA' (ibid.). In this context, it mattered less whether the village of Haçibayram was or was not important in itself – it was, after all, unoccupied – than whether it indicated flaws in the conduct of the EIA. In making this argument, the NGOs drew attention to the part played by the oil company's consultants, Environmental Resources Management (ERM), which had been responsible for carrying out and managing the conduct of environmental and social research in Turkey. It was their fault, and the BTC company's responsibility, that the EIA was so flawed. Additionally the NGOs, which included the London-based Kurdish Human Rights Project, associated the depopulation of Haçibayram with the Kurdish conflict. This last claim was not denied by BTC in Turkey, although the problem was understood not to be a consequence of 'the Kurdish war', as the NGOs claimed, but to stem from the unrest caused by 'some groups' from south-east Anatolia (BTC/EIA 2003: 4-1).[11]

The IFIs dismissed the NGOs' counter-claims. The NGOs had gathered some evidence: that the village was unoccupied for part of the year and that the Muhtar had not been spoken to directly, although his son had. Yet, according to the IFIs, they had failed to gather other pieces of evidence:

> An IFC social development specialist visited the area and found that the villagers do exist, and that they constitute a cohesive community still farming their land through which the pipeline passes, but largely living in a nearby town to avail themselves of facilities and services. They reported satisfaction with the consultation and land acquisition process and compensation amounts paid. (IFC 2003a: 12)

Nor did the NGOs recognise the weaknesses of their own practices: they had only visited the region briefly themselves and were 'superficial, not objective' and 'primarily aimed at discrediting the project's consultation, land acquisition and compensation process' (ibid.: 33). Moreover, according to the BTC company, Haçibayram villagers noted that a group of foreigners had told them that the BTC project would decide compensation values with the villagers, misinforming them of the actual process set out in the RAP: 'the main issue raised by the villagers was about the information they got from a group of foreigners on land valuation methodology ... they asked the reason why DSA [Designated State Authority] used a different approach than what they were told by foreigners' (BTC/EIA 2003: 4-1). Furthermore, although the village was frequently unoccupied, it did still exist as a state entity, with a population that could be consulted. In short, the IFIs reckoned that the campaign's research itself had its own limitations and its own impact that it had failed to acknowledge.[12] Nonetheless, the case of the consultation of the unoccupied village was discussed in the offices of the EBRD in London and at the IFC in Washington. Reports based on the claims of the Baku-Ceyhan Campaign were also published on a number of NGO and news magazine websites.[13] It was an issue that briefly acquired global visibility.

The case of Haçibayram bears out the wider claim made by sociologists of scientific knowledge that empirical evidence is seldom decisive in settling a controversy: 'the problem is that, since experimentation is a matter of skilful practice, it can never be clear whether a scientific experiment has been done sufficiently well to count as a check on the results of a first' (Collins 1985: 2, Collins and Pinch 1993). Harry Collins demonstrated the problem of what he termed 'the experimenter's regress' through studies of laser technology and gravitational radiation research: the 'skilful practice' of public consultation and social impact assessment may be rudimentary in comparison to the technical practice of laser or gravitational research, but the same problem arose.[14] In Haçibayram, the controversy turned on the question of whether particular individuals had, or had not, been telephoned as part of the consultation process, and whether these individuals could or could not represent the interests of Haçibayram. Through their own direct or indirect access to individuals, the oil company, the IFIs, and their NGO critics, all claimed to be able to represent the concerns of the public. By pointing to the case of Haçibayram, NGOs sought to undermine the credibility of the both the EIA and the competence of the company's consultants, by reference to a specific matter of fact. However, this claim was met with the counter-claim by the IFIs that the competence of the NGOs was questionable. Both sets of protagonists in this dispute questioned whether the research of the other had been 'done sufficiently well' or was grounded in 'skilful practice' (cf. Collins and Pinch 1993). In these circumstances, there could be no public resolution to the dispute between the NGOs and the IFIs over the consultation of Haçibayram, although the controversy became less significant once the IFIs' boards had decided to support the BTC project, thereby endorsing their own staff's positive judgement of the oil company's assessment of the consultation of affected communities.

Yet the controversy over the consultation of Haçibayram was more than a dispute about a particular matter of fact, or even just a disagreement about the performance of the EIA, or the decision of the IFI boards to support the development of the pipeline. For the case of Haçibayram could elicit abductive inferences about the conduct of the oil company in general. In this context, the unoccupied houses of the village itself mattered less than its precise position within the series of corridors within which the pipeline would come to be built. Referring to the oil company's own maps of the pipeline corridors James Marriott made this point clear:

> The map shows Haçibayram in the corridor of a new industrial project. For hot on the heels of the East Anatolian Natural Gas pipeline, comes a grander scheme, the Baku-Tbilisi-Ceyhan Oil Pipeline is planned to pass this way. The thin black lines on the white paper are spewed out from the belly of a hard drive. This is the Social Base Line Map, part of the Social Impact Assessment for the Baku-Ceyhan pipeline, produced by ERM: Environmental Resources Management, in London. ERM won the contract from BP in May 2000 to conduct this study. (Marriott and Minio-Paluello 2012: 202, see also Marriott 2003)[15]

In these circumstances, the international NGOs did not occupy a village or a construction site, but rather attempted to disrupt the formation of the space that was so critical to the project. Just as Greenpeace drew others to conclusions about the conduct of Shell through the occupation of the Brent Spar, the Baku-Ceyhan Campaign elicited abductive inferences about the conduct of BP through their multiple interventions in the space within which Haçibayram was placed.

On the basis of their own research in Haçibayram and elsewhere, the coalition of international NGOs sought to speak on behalf of the public in relation to a problem: the failure of the company to meet its obligation to carry out an effective process of public consultation. This could be understood as a form of Deweyian politics, that is, the idea that the formation of a public is necessary in order to address 'issues for which it alone can ensure a settlement' (Marres 2005: 154, Dewey 1927). For the international NGOs, the case of Haçibayram was represented as an issue of public significance, in this Deweyian sense, because the affected communities had the right to expect that they would be properly consulted.

It would be a mistake, however, to view the dispute over the consultation of Haçibayram as simply concerning a single, specific issue. From the point of view of the NGOs, the case of Haçibayram was but one example – generated by their own fieldwork – within a larger campaign against the provision of public support for the construction of the pipeline. While the international NGO campaign was focused on the construction of the BTC pipeline, this campaign was understood by participants as but one element among a number of distinct campaigns that addressed ongoing problems and controversies or, what I have termed, political situations. These political situations revolved around the problem of climate change (Friends of the Earth 2003), the accountability and social responsibility of multinational corporations (The Corner House 2011), and the continuing conflicts over the Kurdish question in Turkey (KHRP 2003). Haçibayram was not just one of a number of disputes along the pipeline, it was also part of a number of other controversies that were not primarily focused on the construction of the pipeline at all. In Gramscian terms, one might view Haçibayram as a particular encounter in a series of overlapping but distinct and evolving 'wars of position' (Gramsci 1971). Conversely, in the view of some IFI staff, the international NGOs' position appeared fundamentally antagonistic to the project, and uninterested in the forms of dialogue that they sought to promote. Their actions, after all, were 'primarily aimed at discrediting the project's consultation' (IFC 2003a: 33). In this view, despite the NGOs' explicit focus on the importance of matters of fact, their factual claims were overdetermined by their antagonistic (Schmittian?) politics and, therefore, could not be trusted (cf. Barry 1999, Mouffe 2000). In short, participants could both experience and address the politics of the dispute of Haçibayram in broadly Deweyian, Gramscian or Schmittian terms.

A Disputed Public

In October 2003, a small demonstration was held outside the EBRD offices on Liverpool Street in London. At the protest, anti-BTC campaigners displayed 'witness statements' by villagers, which reported grievances. These statements, gathered from one or more villages by NGOs, documented complaints over compensation procedures. At the protest, EBRD staff unexpectedly came outside to talk to the campaigners, performing in public, on a London street, their commitment to open dialogue with civil society and stakeholders. However, while in conversation with demonstrators on the Bank steps, EBRD staff refuted the NGOs' claims to be able to speak on behalf of affected communities, citing the lack of public opposition to the pipeline they themselves had observed at a series of multi-stakeholder forums they had organized in Tbilisi, Baku and a number of other major centres along the route of the pipeline, including Borjomi in western Georgia. Later, the EBRD vice-president, Noreen Doyle, wrote of the importance of these meetings:

> As part of the fairness test [that the project would fairly benefit the population], the EBRD worked with the IFC to conduct *the most extensive efforts in the Bank's history to listen to the affected people.* We held public meetings in all the three countries. And we talked with and listened to representatives of many constituencies, including governments and scientists, many special interest groups, and from communities near the pipeline. That gave us a strong sense of support for the pipeline. (EBRD 2004, emphasis added)

What is striking about this exchange on the steps of the Bank is the clash between two genres of public-making. In one, EBRD staff presented the multi-stakeholder forum as a public sphere within which public opinion could be, and had been, freely expressed, issues could be raised, and concerns and criticisms addressed. Meetings such as these had been demanded by Central European Bankwatch, a network of NGOs from across former socialist countries (CEE Bankwatch 2004: 2). In this account, the EBRD and the IFC had created the possibility of informed public debate. The public were expected to present themselves without the mediation of their political representatives or the company: 'the overall purpose was for both institutions to hear *directly from the public* and present their respective Boards with complete and accurate information before making final lending decisions' (IFC/EBRD 2003: 2, emphasis added). During the earlier period of public consultation the company had progressively sought to assemble 'affected communities' through a succession of meetings, 'road shows' and field visits. By contrast, the IFIs attempted to address people in a brief series of public forums. The lack of explicit opposition to the pipeline in the forums led to the abductive inference by the IFIs that there was no objection

Figure 5.1   Witness Statements, outside of the offices of the EBRD, London, 2003. Photo taken by the author

to the construction of the pipeline in general. Indeed, one of the few public critics, the World Wildlife Fund (WWF), itself was criticised by 'local Borjomi representatives'. According to the IFIs: 'WWF had been working in the area for 13 years and these representatives said that WWF had made many promises, such as cleaning up and improving water supply systems in the Borjomi area, but nothing had ever materialized from that' (IFC/EBRD 2003: 45).[16] In short, the capacity of international NGOs to speak on behalf of the affected public was contested by the testimony of other representatives of the public.

But if the multi-stakeholder forums were presented as a genre of public-making that created an open space for the public's concerns to be articulated, some forum participants thought they saw hidden political and economic interests at work. A member of a Georgian environmental NGO reported her sense of puzzlement when, at the IFI meeting in Borjomi, it appeared as if the whole of the population of Borjomi and nearby villages supported the project. However, she claimed to have later learned from a taxi driver that participants had been selected and told to show their support (see also WWF 2003: 15). Elsewhere some informants assumed that those participating in such meetings must have links with political parties, or be close to the government, or were given more time to speak than others because of their status, or were interested in soliciting funds from the BTC company. In Baku, NGOs considered that open opposition to the pipeline, a project considered so central to the interests of the Aliyev government, was evidently risky (CEE Bankwatch 2003). In any case, some observers in Turkey and Azerbaijan argued that local landowners affected by the construction of the pipeline would consider it more appropriate to raise their concerns in private rather than in public forums.[17] These local critics of the spaces of public debate did not ground their judgements about the forums, and the earlier process of public consultation, by conducting systematic empirical political research, but by drawing on their own sense of the local political situation. These observers would agree with the views of critical social scientists who have long recognised that publics can be co-opted or rendered passive by systemic inequalities in economic and political power.

Conclusions: From Affected Communities to Material Impacts

The development of the BTC pipeline posed a huge political problem for both the BTC company and the international financial institutions. On the one hand, given that the project was expected to demonstrate its transparency, its transparency needed to be witnessed by diverse publics. Whereas the Extractive Industries Transparency Initiative was addressed to interested civil society organisations, who could in principle be gathered together in one location, the various publics interpellated by BTC were diverse, poorly defined and geographically dispersed. On the other hand, given that the BTC project was expected to exhibit the values of corporate social and environmental responsibility, it needed to address the needs and concerns of affected communities in particular.

Moreover, the process through which publics were assembled and addressed was itself disputed. For a start, the company and the IFIs' claims to have actually consulted the public or engaged in meaningful dialogue with civil society or stakeholders was contested. The case of Haçibayram suggested the abductive conclusion that the consultation process as a whole was inadequate. In addition, critical observers questioned, in private at least, whether the publics that had been assembled actually were representative of wider publics. The IFIs' claim that they had consulted local publics through 'stakeholder forums' was regarded as especially problematic. Moreover, the idea that there were 'affected communities' appeared necessary, but created problems of its own. After all, it required BTC to define which communities were affected, in order to find out how they thought they would be affected, before it was known whether they were affected or not. In turn the formation of affected communities was likely to have a second-order effect: such communities were likely to be affected simply because they were designated as such. The formation of affected communities appeared to offer a solution to a novel biopolitical problem. However, as we have seen, the form of this solution had further unintended consequences.

The centrality of the idea of the affected community not only provided a catalyst for disputes about the consultation of such communities, it also led to disputes over other issues. For a period of two years, affected communities had been defined in terms of their precise distance from the future route of the pipeline, during which time the company and its consultants had meticulously recorded both their own and stakeholders' concerns about matters ranging from the danger that foreigners might take construction jobs to the preservation of biodiversity in mountainous areas. But as construction work began, the presence of workers, construction materials, lorries and pipes affected the use of land, buildings and the movement of people and animals. As we will see in Chapter 6, in these circumstances conflict increasingly came to revolve around materials and their impacts, and how the impact of materials could be rendered visible or not.

Chapter Six
Visible Impacts

South of Baku the mark of the early history of the oil industry is clearly visible. Rusting oil pumps and pipes litter the polluted landscape. By contrast, BP described the Baku-Tbilisi-Ceyhan (BTC) oil pipeline as 'safe, silent, and unseen' (BP 2004, 2011). In the company's account, the pipeline would be buried, roads would be repaired, and the rural landscape would be restored to its former state. There would be no rusting pipes or rubbish visible on the ground, while the material consequences of the pipeline would flow indirectly through state budgets and community investment programmes. All that would be visible between Baku and Ceyhan would be a few pumping stations, a series of orange and yellow signs marking the route, and a further series of signs warning that the pipeline operated under high pressure and that earthworks should be avoided in its vicinity.[1] One of the critical differences between the older oil industry and a pipeline built according to the new principles of corporate social and environmental responsibility was that the newer form should literally disappear from view:[2]

> Pipelines are the safest, most secure method of transporting crude oil – particularly when buried underground as the BTC pipeline is along the entire route. The only installations above the ground will be eight pumping and/or metering stations and unobtrusive block valves every 20 km or so. All will be designed to cause *minimal intrusion* into the landscape and the lives of local communities ... Once the pipeline has been buried *reinstatement* will ensure that the landscape returns as closely as possible to its original

Material Politics: Disputes Along the Pipeline, First Edition. Andrew Barry.
© 2013 John Wiley & Sons, Ltd. Published 2013 by John Wiley & Sons, Ltd.

condition – leaving *little if any evidence* of the pipeline's presence for the remainder of its 40-year operational life. (BTC 2003c: 7 and 21, emphases added)

If the pipeline was going to be physically invisible, then its impact on the environment would be made visible through the production of copious quantities of information. In other words, its invisibility – lying beneath the surface, and not disturbing its environment – had to be demonstrable and made public. Reinstatement did not just have to occur, but had to be shown to have occurred. Indeed, a panel of senior industry and government figures set up by BP to advise senior management about the development became particularly concerned about the 'real problem' in getting the Turkish company, Botaş, and the Turkish government to address the question of reinstatement (BTC/CDAP 2004: 5, see also BTC/CDAP 2007)[3], and the question of the quality of reinstatement remained a critical issue for the lenders group independent environmental consultant (e.g. BTC/IEC 2010). The invisibility of the pipeline – expressed in the ideal of reinstatement – would itself have to conform to international standards.[4] The intrusion of the pipeline into the landscape and the lives of local communities needed to be assessed so that it could be minimised. Evidence would

Figure 6.1   Visible signs. Photo taken by author

need to be collected about the pipeline's presence throughout its life in order to check that there was little, if any, evidence of its presence. In this way, the construction of the BTC pipeline was expected to embody a core principle of transparency and corporate responsibility: that its impacts should become visibly invisible.

In this chapter I focus on the idea of impact, and the emergence of disputes over the question of what the impact of pipeline construction had been or might be in the future. This focus on impact directs us towards the relationship between two distinct ways of thinking about the relations between politics, materials and the environment. On the one hand, as I have suggested, the idea of impact has become critical to the discourse and practice of a responsible corporation. Negative environmental impacts are expected to be anticipated, minimised or reduced through a variety of what Bulkeley and Watson, following Foucault, call 'modes of governing' that include, for example, disposal, diversion, eco-efficiency and recycling (Bulkeley and Watson 2007: 2740, see also Lemos and Agrawal 2006, Bridge and Perreault 2009). Reinstatement can be seen as another environmental mode of governing in these terms. On the other hand, the notion of impact points to a very different approach to the study of environmental government that dwells not so much on the discourse or practice of modes of governing, but on the materiality of the environment that it is intended will be managed. Impact implies the existence of a material force that has a physical effect on something else, such as the impact of aircraft noise on a town, or the impact of chemical pollution from a metallurgical factory on a nearby farm (cf. Callon 1998b: 245). This latter sense of the term impact directs us towards an interest in the nature of the materiality and force of things that have an impact, as well as the persons and materials that are impacted upon (Bennett 2005, Braun 2008b, Braun and Whatmore 2010, Gregson and Crang 2010: 1028).

In this chapter I develop two arguments that draw together these two contrasting accounts of impact: one concerns environmental impact as an object of modes of governing; the other concerns the materiality of things that are the cause or consequence of impacts. I argue, first of all, for the importance of distinguishing between two ways in which impacts can be understood. One way of conceptualising impact is in the terms of what I will call a *measured impact* (cf. Barry 2005).[5] This idea points to how environmental problems are constituted as 'impacts' through multiple methods and techniques, including modelling, sampling, epidemiological studies and direct observation (Callon 1998b: 259, Latour 1999: 24–79, Landström et al. 2011). Scientists concerned with the problem of environmental impacts do not aim to grasp such issues in all of their complexity; their work is expected to enact impacts in forms that render them amenable to management. Impacts do not exist independently of their enactment. Rather, by understanding an event such as traffic noise or river pollution as a (possible) 'impact', it is given a new existence. It becomes something about which information should be produced, in order to determine whether it should be treated as an impact or not. In this way, scientific research on environmental impacts can be likened to a form of medical investigation concerned with determining the nature of a disease in the patient's body. The medical doctor does not attempt to understand the body as a totality but, working with other specialists and deploying a variety of techniques, multiplies the forms of the existence of the disease (Mol 2002: 119). Likewise, the environmental scientist concerned with the problem of the impacts of construction work does not need to conceptualise the environment as a totality or system; rather, the environment is abstracted in a diversity of informational forms, as 'impacts', in order to determine who or what is responsible for them. Nonetheless, the abstraction of environmental impact should not be understood as a projection of scientific categories onto nature. Rather, impacts are abstractions. As A.N. Whitehead argued, 'abstraction expresses nature's mode of interaction and is not merely mental. When it abstracts, thought is merely conforming to nature' (Whitehead 1927: 25–26, Halewood 2011: 147–148).

However, the idea that impacts are abstractions can be contrasted with the idea of what Whitehead termed an event. If impacts are understood as events, then we need to attend to the ways in which impacts often have many causes and contain multiple elements, which cannot be rendered into objects of scientific analysis, for 'it requires no illustration to assure you that an event is a complex fact, and the relations between two events form an almost impenetrable maze' (Whitehead 1920: 78). The notion of the event directs us towards the 'impenetrable maze' of relations that could enter into any particular impact, including those that cannot readily be measured or assessed but are discernable nonetheless. The anger and frustration of villagers at the lack of response of the BTC company or the World Bank to a grievance, for example, is an affective impact, which is not generally recorded. At the same time, the assessments of environmental impacts that help to constitute what I have termed measured impacts are likely themselves to have a further impact that could be difficult to determine.

The second argument developed in this chapter concerns the space constituted by the impact of the construction and operation of the pipeline. In the last chapter we saw how the oil company established a series of 'corridors', the width of which was determined at an early stage in the project. Impacts, however, cannot fully be determined in advance, even if some of them may be anticipated. Specific impacts may be unexpected, and their nature and extent is often uncertain. They are the emergent effect of particular circumstances. Equally importantly, we cannot assume that the space given by the existence of various impacts corresponds to the width of the pipeline corridors. On the one hand, many things existing within the pipeline corridors, such as polluted streams or damaged buildings, which could in principle be considered to be the impact of construction work, may have predated the construction of the pipeline. On the other hand, the impact of pipeline construction work is likely to go beyond the limits of the pipeline corridors. While the informational space of the pipeline corridors maps onto a narrow strip of land, the space of impact projects a more complex topology (Murdoch 2006, Blok 2010). Not everything that occurs within the pipeline corridors can be enacted as a measured impact, but some things that occur beyond the pipeline corridors could be. In this way, the assessment of impacts folds a series of dispersed and specific materials together into a heterogeneous and shifting field of events. While 'affected communities' were thought to exist within a well-defined corridor, impacts could not necessarily be contained in this way. As we shall see, the difference and interference between these two spatial forms contributed on occasion to the emergence of disputes.

Impacts can take diverse forms and they can be localised or dispersed. In the development of the BTC pipeline, it was particular cracks or sets of cracks in buildings, pipes and walls that came to have remarkable transnational significance. Indeed, throughout my fieldwork, environmentalists, pipeline workers, oil company employees and villagers frequently drew my attention to the existence of specific cracks, gave me photographs of cracks, or encouraged me to drive to see cracks and take photographs of them myself. After all, the evident presence of a crack cannot simply be explained away as an ideological projection or the product of a fertile imagination (Barry 2002). The presence of cracks seemed, for many observers, to lead to abductive conclusions: they pointed to the existence of larger forces that had caused them to occur. Caroline Humphrey notes that in Russia, the presence of surface contamination and rusting equipment and pipes can be experienced as the all-too-obvious consequences of the decline of the post-Soviet political system (Humphrey 2003). In the case of the BTC pipeline, cracks were taken to be clues that pointed to the existence of defects in the ethical project of which they were a part.

In this chapter I focus on two contrasting instances, both in Georgia, in which the significance of cracks in buildings became matters of dispute. One set of cracks occurred in houses and walls in Dgvari in the Lesser Caucasus, a village that became the focus of an extraordinary level of transnational interest over several years. The other existed in a number of houses along the side of a road in the village of Sagrasheni, not far from Tbilisi. Both sets of cracks were drawn to the attention of the ombudsman of the IFC in Washington, DC, and both were investigated by consultants working for the BTC company, as well as by the Georgian NGO, Green Alternative. These contrasting cases point to how the assessment of impact contributes to the formation of a contested topology. At the same time, they indicate the complexity of the ways in which the behaviour of materials is mediated through the assessment of impact.

Landslides

In the process of making an episode for a short series of documentary films on the oil industry, a crew from a British independent television company visited the village of Dgvari, near the edge of the Borjomi-Kharagauli National Park in south-west Georgia, accompanied by a member of the Georgian environmental NGO, Green Alternative. The group was watched by members of a different NGO that their guide claimed had been set up to support the construction of the pipeline (Cran 2005). One of the producers of the documentary described the visit in the following way:

> The NGO people [from the second NGO] watched us as we interviewed villagers whose houses were literally sliding down the valley in this geologically unstable area of Georgia. Many villagers think that the pipeline will make the landslides worse, and some were clearly disconcerted by the presence of the dozen or so other visitors. [Their guide from Green Alternative] also seemed somewhat intimidated by them ... When we got back to the hotel, we too started to fell intimated. There to greet us was a group of people from the NGO, with a camera crew and the local governor. They demanded to know who we were, and the governor asked us for our papers. He went on to accuse our fixer ... of being against the pipeline, and told us that we would come to regret what we had done that day. (BBC 2005)

The visit of the documentary film-makers to Dgvari was brief, but their experience is nonetheless instructive. Despite the presence of the Georgian environmental NGO and the brief mention of the occurrence of landslides, the television company did not present the problem of the village primarily as an environmental one.[6] In the logic of the film, the village was situated in the midst of a struggle for power between progressive elements of civil society and shadowy elements of the state. The focus of the British film-makers' interest was on the relationship between the two Georgian NGOs and the villagers, and the sense of tension and threat that they generated. In one sequence, the men who were observing the film-makers as they talked to villagers can be seen in the background. According to the film, these were the circumstances within which the pipeline would be constructed and to which it would contribute in the future. Viewed in this way, the conflict in Dgvari was but one element of a wider conflict between oppositional parties: civil society and media organisations on the one hand, and the Georgian state on the other (Hamilton 2004, Manning 2007). The occurrence of landslides was not something that needed further analysis in this context; it merely contributed to a sense of the instability of the setting and the tragedy of the village. In other sequences, further threats to the pipeline were sketched, including the potential for renewed warfare in Nagorno-Karabakh, and the ongoing conflict between the Turkish army and the Kurdistan Workers' Party (PKK) in south-eastern Turkey. In effect, the environmental disturbance of a village became a synecdoche to describe the instability of the region as a whole. The documentary did not provide an account of the environmental impact of pipeline construction, but developed a general sense of an atmosphere of intimidation and instability, reflecting wider Euro-American narratives about the region in this period (cf. Dodds 2003).

The British documentary film-makers were not, however, the only outsiders to come to the village of Dgvari. Others who came to the valley included: social researchers from consultancies and international financial institutions (BTC/SRAP 2004b: C-43; BTC/SRAP 2005b: C-46; IFC/ CAO 2004a&b, 2008); geoscientists; archaeologists (BTC/ESAP 2002b: 96); officials; journalists from the US (Rondeaux 2004), the UK, Australia and Denmark[7]; international NGO 'fact-finding missions' (Bank Information Center et al. 2003, Center for Civic Initiatives et al. 2005a); a Georgian documentary film-maker supported by the French television channel Arte (Kirtadze 2005); an American photojournalist who published her work in *National Geographic* (Rivkin 2011); and myself. No doubt many of these visitors were shown the same cracks – that run down the walls and across the terraces of many of the village houses – as I was shown when visiting Dgvari together with a member of the Georgian Young Lawyers Association (GYLA). These cracks were repeatedly observed over several years, and accounts of their existence were circulated worldwide, in reports and political pamphlets, on websites and on film. For a period, this tiny village of approximately 115 households became a site of global interest. Its landslides and cracks became objects of what John Urry once termed the 'tourist gaze' (Urry 1995).

According to international NGOs, the residents of Dgvari believed that the construction of the pipeline could increase the severity of the landslides in the village (Bank Information Center et al. 2003: 6, Green Alternative et al. 2004). The BTC company and the IFIs argued, however, that there was little relation between the construction of the pipeline and the occurrence of landslides. It was true that Dgvari was suffering from increasingly frequent landslides that were progressively rendering its houses unsafe, and it had already been acknowledged by the Georgian government that the village should be moved as a result. Indeed, in 2003, geologists from the Georgian State Department of Geology had completed an initial assessment of the civil engineering and geo-dynamics of the village and confirmed, according to the IFC, that 'it was not possible to stabilize the landslide area and that Dgvari should be relocated to a safer location' (IFC/CAO 2004a: 5). The IFC felt that the condition of the village was a matter for the government and not, therefore, a problem for which the oil company should take responsibility (IFC 2003a: 31–32). One informant working in the valley reckoned the landslides might have been exacerbated by logging, but they were certainly not caused by construction work. The problem is one familiar to environmental geographers (Guthrie 2002, Goudie 2006). The villagers were only approaching the oil company, in this man's view, because the company actually might be persuaded to address the problem, while the government had not acted yet and still might not act in the future. Moreover, geoscientists paid by the BTC company confirmed that there was no relation between the pipeline and the likelihood of landslides (BTC/SR 2005: 11, cf. BTC/SRAP 2004b: C-43). They agreed that the problem facing the villagers was critical, taking numerous photographs and detailing the number of houses already suffering serious damage including cracking. Whereas the documentary film-makers, international NGOs and Georgian environmentalists drew attention to the imminent arrival of the oil company, the geoscientists recognised that there were longer-term processes at work. As a later report by a team of Georgian geoscientists noted 'the landslide territory [of Dgvari] unites deformation zones of quite different generation, age, mechanism, kinematics and depth deposition' (Tatashidze et al. 2006). Certainly landslides had occurred in the valley even before the village was built.

In suffering from landslides, Dgvari was not at all unusual: throughout the region the landscape was gradually being forced upwards, and other villages had had to be resettled during the Soviet period. Landslides were most likely to occur not because of the construction of the pipeline, but because of periods of high rainfall or earthquake activity (Shiston et al. 2005, Lee and Charman 2005: 99, see also BTC/ESIA: 2002g: 13). Human activity could also have contributed, primarily on account of leaking water pipes and the discharge of domestic water and sewage. The BTC study attended to the interaction of society and the physical environment, linking a geoscientific analysis of the landslide system to an account of the stability of Dgvari's houses. The analysis had to be based on the judgement, knowledge and experience of the scientists, grounded not just on their observations in the village and the valley in which it was situated but on the limited 'information about the possibility of landslide triggering events across Georgia' (Lee 2009: 454). On the basis of the report, as well as the earlier work of the Georgian scientists, it was acknowledged that a long-term solution to the problem would almost certainly involve relocation (IFC/CAO 2004a: 5–6, BTC/SRAP 2004b: C-43). Nonetheless, landslide risk assessment could not offer an explanation as to why the villagers sought to appeal to the oil company rather than the government. Nor would it have involved any examination of why the valley was economically impoverished or why its infrastructure was so poor. Nor was the study concerned with the relations between the villagers and the residents of other villages in the valley. Green Alternative welcomed the confirmation that Dgvari was suffering from landslides but, given the absence of any account of pipeline construction, questioned the consultants' conclusion that construction of the BTC pipeline and its subsequent operation would not have any impact on the landslide system (Green Alternative 2004a, BTC/SR 2005: 11). Moreover, according to the environmentalists, the earlier Georgian government study of Dgvari was never completed, as the project had been terminated due to 'lack of funds' (Green Alternative et al. 2004: 18). The villagers themselves received a letter from the Georgian International Oil Company (GIOC) summarising the conclusions of the BTC report (BTC/SRAP 2004b: C-43).

But to understand why Dgvari should have attracted so much interest it is not sufficient to know about its relation to the landslide system. We also need to understand its specific relation to the route of the pipeline. The village lies near the end of the valley, on the other side from the pipeline, which ran much closer to the adjoining village of Tadzrisi, and Sakire, a village located higher up on the opposite slope to Dgvari. Yet despite its proximity to the route, the village had been omitted from the Environmental and Social Impact Assessment and Resettlement Action Plan. While the impact of pipeline construction on Tadzrisi and Sakire had had been assessed, and the residents of Tadzrisi and Sakire had been consulted face to face, the impact on Dgvari had not been assessed and the residents had not been directly consulted (BTC/RAP 2002a: V-2, Green Alternative/CEE Bankwatch 2005: 39). Moreover the location of Dgvari was also not given by BTC on a map of 'settlements along the pipeline route'. It was this omission that provided the basis for a complaint by the villagers and Green Alternative to the World Bank ombudsman: their claim was that 'unfortunately the sponsor "forgot" to study the impact of BTC construction on this village' (Green Alternative 2004a: 1–2, CEE Bankwatch 2006: 41–42). In practice Dgvari had figured in the BTC ESIA but only as a potential site of 'cultural heritage' (BTC/ESAP 2002b: 96) that was said to include burial mounds and pit tombs from the Middle Bronze Age (BTC/ESIA 2002f: 14).

Whatever its direct environmental impact might have been, the construction of the pipeline did affect the village profoundly, given its specific relation to the corridors within which the pipeline would be situated. This had at least three aspects. First, as Green Alternative argued, Dgvari was located within 2 km of the pipeline route, and it should have been consulted as an 'affected community' or included within the environmental and social impact assessment report. Secondly, the villagers mistrusted the findings of the BTC consultants' study (IFC/CAO 2008: 5–6). After all, 'the 44 m construction corridor crossed one of the landslide zones' near the village and other landslide sections were only 150 m from the corridor (Green Alternative 2004a: 3). Yet, according to Green Alternative, the BTC consultants' report had only briefly considered the impact of pipeline construction on the prevalence of landslides (CEE Bankwatch 2006: 42), and had not considered the view of a Georgian geologist that the landslides might have an impact on the pipeline (IFC/CAO 2008: 5–6). Thirdly, while some of the residents of the neighbouring village of Tadzrisi had been

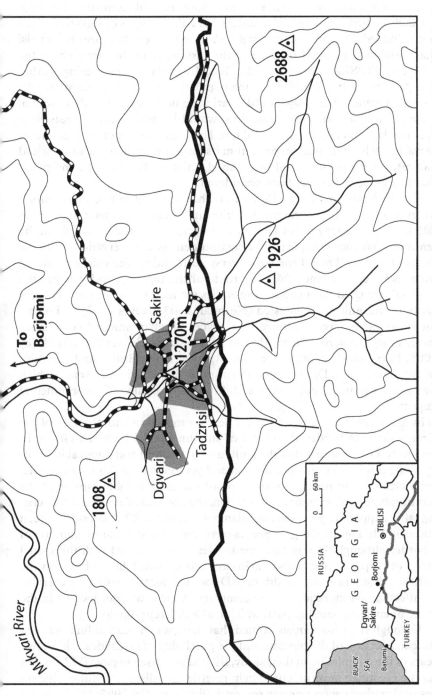

Figure 6.2 Map of Dgvari and surrounding villages. Map prepared by Ailsa Allen, School of Geography and the Environment. Reproduced by permission of the University of Oxford

eligible for substantial compensation from the oil company, for both privately owned and communal pasture land, the villagers of Dgvari received none. Indeed, the route of the pipeline had to be brought nearer to Tadzrisi than originally planned because of the instability of the slope above the village (BTC/SRAP 2008, see also NCEIA 2004: 4), generating further compensation.[8] In a valley where the population was largely reliant on subsistence farming, the villagers of Dgvari were no doubt aware that some of their neighbours' financial difficulties would be addressed, whereas their apparently greater problems would not. The social impact of the pipeline on Dgvari may have derived, in part, from its impact elsewhere (in Tadzrisi and Sakire); yet the possibility of this second-order effect, and the atmosphere that it generated, had not been considered.

Kay Anderson and Bruce Braun note that 'it's hard to know anymore, if we ever did, where environment begins and ends' (Anderson and Braun 2010: xi). In thinking about Dgvari, we can read this observation about the limits of what counts as part of the environment as an observation about the limits of impact. Part of the controversy surrounding the village stemmed from its precise distance from the route of the pipeline, and whether the 'impact' stretched this far. Social scientists are used to the idea that borders are both dynamic and contested (Newman 2003, Reeves 2008) and political geographers have examined how spaces of environmental governance increasingly cut across the borders of nation states (Bakker 2003, Bulkeley 2005, Barry 2001, 2006, Lemos and Agrawal 2006, Bridge and Perreault 2009). In this case, Dgvari lay on the disputed boundaries of a space, which was not a territory, but rather a shifting field of sites of environmental impact.

In this respect, the case of Dgvari bears comparison with the case of the Greek-Albanian border, the object of an ethnography by Sarah Green. In her study, Green argues that the fault lines of political and physical instability along this border are both distinct and yet interwoven. This 'can be seen most clearly', she notes, 'through one of the most notable aspects of the geomorphological accounts of the region: the absence of any detail concerning what happens beyond the Albanian border, on the Albanian side' (Green 2005: 96). In Dgvari, the geosciences were also critical to the constitution of a border although, in this case, the border was associated with the spatial limits of corporate social and environmental responsibility rather than the territorial borders of states. Although Dgvari was located within the corridor of 'affected communities', in the company's view it was not located in the space of environmental impact, which had a different topology.

Although the geoscientists argued that there was no causal link between the construction of the pipeline and the possibility of landslides, within two years of the completion of their study, BP had promised to pay US$1 million via the Georgian government to help resettle the village as a 'humanitarian gesture', without accepting any responsibility (BTC/SR 2005: 11). How did

Figure 6.3   The impact of landslides, Dgvari, September 2010. Photo taken by author

this sequence of events happen? It is not possible to say, since the content of negotiations between BP and the government were not published , and BP's fund was refused by the Georgian government in any case, who offered the residents of Dgvari 800,000GEL (≈ US$400,000) instead (BP 2005: 13, BTC/SRAP 2005b: C-45, BTC/SRAP 2006a: C-13, Pagnamenta 2005). Nonetheless, even after this formal resolution of the controversy of Dgvari, and three years after the visit of the documentary film-makers, journalists were still regularly attracted to the village, reporting 'that villagers say pipeline excavations have seriously destabilised surrounding lands and allege that promised amounts of compensation have not been paid' (Cooke 2006). Although many of the residents of Dgvari had moved elsewhere by 2010, on my return to the village, some still remained living in houses that were progressively collapsing around them.

Despite the claim by the geoscientists that the construction of the pipeline had no environmental impact on the village, it is quite evident that it did have an impact of a kind. However, to understand the event of Dgvari it is necessary not to confine analysis to the landslide system that undermined the stability of the village houses: 'it is the business of rational thought to describe the more concrete fact from which that abstraction [the landslide system] is derivable' (Whitehead 1933: 186). As we have seen, such an analysis would also need to address a series of contingent processes that were specific to this village at this time. These include not just the nature of the landslide system in the valley, the prevalence of logging, the deteriorating state of village houses, and the poverty of their residents, but also the dynamics of the political situation that had emerged around the question of the relation between pipeline and landslides in the Borjomi region more broadly (Chapter 2). Most of all, it is necessary to attend to the very specific location of the village to the corridors that had been forged along the length of the pipeline, which structured the limits of corporate responsibility and transparency (Chapters 4 and 5). The extraordinary level of international interest in the village Dgvari derived from the contingent and ambiguous position of this village within this complex, shifting, contested and affected space.

There is a resonance here with the work of Doreen Massey on the politics of place. For Massey, there is something of a natural affinity between places and politics: 'attempts to write about the uniqueness of place have sometimes been castigated for depoliticisation. Uniqueness meant that one could not reach for eternal rules. But "politics" in part precisely lies in not being able to reach for that kind of rule; a world which demands the ethics and responsibility of facing up to the event; where the situation is unprecedented and the future is open' (Massey 2005: 141). In this way, for Massey, place is an event in a double sense. On the one hand, what is unique about a place is the way that, however temporarily, it acts as a site of dynamic interference or relation, of diverse elements, human and non-human. Places, such as Dgvari, generate effects and affects that are more than simply the sum of elements that they draw together and transform. On the other hand, a place is an event in the sense that it does not have a given coherence or fixed identity or scale, but is multiple and in process: 'by sharp contrast to the view of place as settled and pre-given, with a coherence only to be disturbed by "external" forces' (ibid.). In relation to Dgvari, the question of whether forces were 'external' or 'internal' to the place, as well as the disturbances that these forces generated, were themselves a matter of dispute.

Vibration

The space of impact is not a continuous territory or corridor with well-defined borders. Impacts may be located in specific places, or spread over large areas. They are associated with a whole series of different objects and processes, including noise, waste, exhaust fumes, emissions, sewage, vehicles, vibration, concrete, sand and dust. The generation of events that might have to be considered and managed as impacts can certainly be anticipated through environmental impact assessment, but such events cannot be avoided altogether. The space of impacts is not fixed: it is a shifting field of events, some of which may be more likely, more extensive, or longer lasting than others:

> A priority throughout project planning, design and the ESIA process has been the avoidance of potential environmental and social impacts. This has resulted in many design modifications and a suite of measures ... that will avoid many potential impacts. There remain impacts that are either likely to occur or are unavoidable. For these impacts mitigation measures have been developed to minimise the likelihood, extent or duration of occurrence. (BTC/ESAP 2003: 11)

Company documents describe a whole series of ways of limiting, tracing and mitigating impacts. Impacts may be unavoidable, but they can nonetheless be governed. Employees can be trained to follow correct procedures when driving, operating machinery and disposing of waste. The emissions of various gases including nitrogen oxides (NO_x) and sulphur dioxide (SO_2) can be regularly monitored, as can the levels of pollutants in rivers and streams. Oil spills resulting from illegal tapping can be monitored and addressed. Local residents can be made aware of likely hazards, such as the presence of ditches. As Bulkeley and Watson suggest, studies of governmentality have tended to focus on the government of human actors. However, any analysis of environmental impact must address how the activity of materials as well as the activity of persons is governed (Bulkeley and Watson 2007, Foucault 2007).

Vibration from construction traffic is a rather minor source of impact in comparison to many others (cf. Morris and Therivel 2009). Indeed, in their 'Contractor Control Plan' for Georgia, the company was primarily concerned with the potential impact of traffic vibration on wildlife and fauna, although it was acknowledged that noise, vibration and dust from traffic was a nuisance to residents (BTC/ESAP 2003: 11). Nonetheless, as construction progressed across Georgia in 2004, complaints about the impact of vibration proliferated. In their 2005 report, for example, specialists contracted by the oil company to assess its compliance with their environmental and social commitments noted that a large number of complaints had been made about construction traffic in particular: 'nearly 75 percent of the rejected grievances [i.e. 45 percent of the total number of grievances] are related to allegations of cracks appearing in houses due to construction traffic vibration, or to bee hive owners who claim compensation' (BTC/ SRAP 2005b: C-14).[9] An earlier report by an independent environmental consultant reporting to the project lender group observed that 'aside from the issue of orphaned land, *most grievances* have related to claims of construction noise and damage to houses caused by vibrations from blasting and trucks' (BTC/IEC 2004: 54, emphasis added). In these circumstances, far from being a minor impact, vibration came to have a remarkable significance that drew the attention of international financial institutions and NGOs as well as community liaison officers, environmental specialists and engineers. I return to consider the interest in the impact of pipeline construction on beehive owners in Chapter 8.

One reason that vibration is likely to have caused 'most grievances' was that construction traffic moved across an area that went far beyond the very limited territory defined by the pipeline construction corridors, and a substantial number of people could therefore potentially be the object of an impact from construction traffic that was not contained in these corridors. Moreover, given that most roads in the vicinity of the pipeline were unmade and uneven, passing construction traffic inevitably caused vibration in nearby houses. At the same time, traffic vibration was experienced somatically and immediately. Vibration was felt, and generated an intensity of feeling. The prevalence of vibration from traffic demonstrated that the pipeline had an impact on villages that were not considered to be 'affected communities'. Consider the following complaint made by the residents of the village of Sagrasheni, supported by the Georgian NGOs Green Alternative and GYLA:

The eight family [sic] of Sagrasheni village have the same problem [damage] from the heavy trucks movement. The heavy trucks that transport construction material are endangering the houses situated near the road, especially those houses whose support walls border the roadside. These trucks are moving along the road from early in the morning until late in the evening. The vibrations caused by the heavy traffic are shaking the buildings, whose walls are cracking. (Green Alternative 2004a: 7)

The movement of construction traffic along the road through Sagrasheni was not just experienced by village residents, but also monitored by the BTC company, whose local Georgian Community Liaison Officer took photographs of construction traffic while also passing on the villagers' concerns to her managers.[10] Similar complaints about the damage caused by heavy construction traffic were raised in the village of Atskuri further to the west, whose residents were also said to be concerned about the damage caused to the village's fortresses and fourteenth-century Cathedral of the Mother of God (CEE Bankwatch 2006: 74–77).

Vibration from traffic was clearly and immediately felt and seen by the residents of Sagrasheni and evidently had an affective impact, but there was still a question of whether this could be categorised as 'impact'.[11] In response to the complaints, three pieces of field research were carried out in Sagrasheni and other villagers. One field experiment was carried out by specialists from the principal contractors, who simulated the impact of construction traffic by driving vehicles at a steady speed along the route and recording the resulting vibration in the walls of village houses using specialised monitoring equipment. The contractors performed similar tests in Atskuri and Vale as well as Sagrasheni. These tests were said to show that the vibration levels were 'significantly below internationally accepted standards of what could cause structural damage to buildings' (IFC/CAO 2004a: 7). However, the Sagrasheni villagers disputed whether sufficient numbers of tests had been done, whether the trucks used in the tests were representative of the trucks that actually went along the road, and whether the tests were scientific:

> When villagers demanded compensation for the structural damage, SPJV argued the cracks were a result of time and general degradation. [The contractors] manipulated a 'vibration test' by driving an empty construction truck past the homes and claiming that the measured vibrations were not strong enough to destroy a house. Yet vehicles passing through Sagrasheni are mostly heavily loaded. (Centre for Civic Initiatives 2005a: 19)

As we saw in the case of Haçibayram, the credibility of particular empirical claims could be undermined through an account of the conduct of the test, and the skilful practice of those who performed it. Here, it was the villagers themselves that questioned the skilful practice of the specialists (cf. Wynne 1996). A second field visit was from the World Bank ombudsman's office in July 2004, in response to the complaint by Green Alternative. In their report, the ombudsman's office reckoned that the response of BTC and its contractors was 'appropriate' (IFC/CAO 2004a: 8) but that the vibration tests should be carried out again in a different form: 'using a *transparent, consensual and participatory* methodology that would include independent monitoring by Complainants and NGOs' (ibid.: 7, emphasis added). For the ombudsman, measurements of vibration had to be not merely accurate, but also transparent and socially robust (Nowotny et al. 2001, Barry et al. 2008). A participatory methodology provided the ombudsman's solution to the problem of finding a way of determining the quality of the tests independently of the output of the tests themselves (cf. Collins 1985: 84). No indication was given, however, as to how measurements of vibration would be carried out in a participatory manner, or how villagers or NGOs would themselves monitor vibration independently of BTC or their contractors. The question of how one might design a participatory forum in which the residents themselves participated in the production of knowledge was raised but not addressed (cf. Callon et al. 2001). Instead, the issue of the Sagrasheni cracks was taken by international NGOs to be another indicator of the gap between the claims of BP and its environmental performance in practice (Centre for Civic Initiatives et al. 2005a: 15).

  In August 2005, however, BTC commissioned a further study of Sagrasheni and other villages that had made complaints to the IFC ombudsman about the impact of traffic vibration (IFC/CAO 2008: 7–9). As a field researcher, the engineer who carried out the BTC study would have been attentive to the specificity of the site, its materials and their relations. He may have listened to the villagers' stories of the sequence of events that had led to the emergence of the cracks and the performance of the earlier vibration tests by the contractors' engineers. Yet, although the engineer did not carry out further tests he did not necessarily need to pay too much attention to the villagers' account of their experience of vibration.[12] After all, as earlier research had demonstrated, the human body was much more sensitive to vibration than house walls, even if they were in poor condition, as they were in Sagrasheni (Transport and Road Research Laboratory 1990, Hume 1995). Walls are able to absorb levels of vibration that people would find uncomfortable. As a result, the affective experience of the body provides a poor guide to the likely impact of vibration on walls.[13]

How then, if he did not conduct new tests or listen to the villagers, could the engineer possibly determine whether the cracks should be considered impacts or not? Certainly, there were reasons to doubt that the tests carried out in Sagrasheni and elsewhere were comparable to tests previously performed in Western Europe. Indeed, the report concluded, as BP acknowledged, that there were 'shortcomings in the adequacy of [vibration] monitoring when subject to international scrutiny' (EBRD/IRM 2009: 14). In their account of the dispute, Green Alternative went further, documenting the key points of the engineer's argument even though his conclusions ultimately refuted the environmentalists' and residents' claims. In particular, the engineer noted that those who performed the earlier set of tests had not had specialist training and vibration monitoring devices should have been re-calibrated at the time of the tests. Moreover, the contractors' engineers had incorrectly assumed that the floor or the windowsill closest to the cracks was the right place to measure vibration levels, whereas British and German standards recommended that monitoring devices should be placed at the foundations of buildings. There was also limited assurance as to whether the weight and speed of vehicles used in the test corresponded to actual conditions (CEE Bankwatch 2006: 68).

Although there had been shortcomings in the skilful practice of the earlier tests 'when subject to international scrutiny', the results of the contractors' tests were nonetheless consistent with what might have been expected on the basis of the results of tests carried out in recognised international laboratories. The answer to the question of whether vibration had an impact did not depend on the observations of the village residents, or on the conduct of a transparent, consensual participatory process that was initially recommended by the World Bank ombudsman (IFC/CAO 2004a: 7), but upon the comparability of the tests performed by the contractors in the Georgian villages to those performed in recognised laboratories in Western Europe. The IFC ombudsman accepted the engineer's conclusion and closed the case in June 2006, while residents received no compensation (Bretton

Woods 2006, IFC/CAO 2008). Although it was acknowledged the traffic management plan may not have been strictly adhered to, the BTC study concluded that it could not be demonstrated that the houses of Sagrasheni and other villages were affected by construction traffic. In short, it was unlikely that construction traffic had been the main cause of the cracks in the village houses. Two years later, BTC advisors reported that the 'complainants viewed positively BTC Co.'s efforts in engaging independent experts to conduct vibration testing, but were disinclined to accept the results: "*the problems have not been resolved, but we have tended to forget with the passage of time*"' (BTC/SRAP 2007: C-13, emphasis in original).

Although it appears to be a straightforward case, the dispute over the cracks of Sagrasheni illustrates some of the complexity of impact and its spatiality and temporality. On the one hand, the space of impacts (and potential impacts) includes a whole series of dispersed events, scattered primarily along the route of the pipeline and along the routes taken by construction traffic. This space is not a territory, or a network, but can best be described as a shifting and contested field of events. On the other hand, these events, such as the cracks in the houses of Sagrasheni, could only become what I have termed 'measured impacts' by being comparable with experiments and tests elsewhere, such as those conducted previously in the Transport and Road Research Laboratory; in this way they become part of a transnational metrological zone (Barry 2002, 2006). It is in the context that the inequality between the resources of the company and of the villagers becomes particularly significant. For in the absence of well-financed public scientific institutions in Georgia, it was only the company and its contractors that had the resources to establish whether the results of tests were, or were not, comparable to those that had previously been performed in laboratories in the UK, Germany and elsewhere. It was the standardised and experimental walls and trucks tested in these Western European laboratories, rather than the somatic experience of residents, that provided the basis on which events could be judged to be (measurable) impacts.

Yet if this apparently minor dispute over the sources of cracks tells us something about the topology of impact and its political economy, it also tells us something about the structure of the archive of documents made public by BP, and the limits of the disputes that they made possible. After all, the dispute between the residents of Sagrasheni and the BTC company focused on the technical question of the impact of vibration on houses. Although the villagers questioned why traffic travelled at such a speed along this road, this was not a key issue for engineers, who were concerned with the physical buildings themselves, or the IFC ombudsman who proposed the possibility of a participatory methodology. During the same period an independent environmental consultant, monitoring the project on behalf of lenders, raised the question of the relation between the BTC company its contractors, and recommended that BTC needed to improve its control of the procedures for environmental monitoring by contractors (BTC/IEC 2004: 23), while one BTC employee noted that 'the reputation protection tendency of the construction company is generally lower than that of the operator' (Spence 2004: 3). While the studies carried out in Sagrasheni and Atskuri focused on the immediate causes of cracks, they were not concerned with the organisation of the industrial process that might or might not have contributed to these possible impacts. The case of Sagrasheni points to the importance of the distinction between the visibility of environmental impacts, which were subject to both political fieldwork and 'external monitoring', and the relative invisibility of industrial practices, which were largely only subject to 'internal monitoring' (cf. MacIntosh and Quattrone 2010, Callon 1998b: 257). Thus a division exists between what is made public, which includes the field of possible impacts, and what is not, which includes the field of industrial organisation. Environmental impact is a form of abstraction: it both simplifies and adds to what already exists. It is part of what, following Alberto Toscano, we might further call the 'culture of abstraction' of responsible capitalism (Toscano 2008, Halewood 2011: 147–170).

Impact and Action

Can we take the disputes in Dgvari and Sagrasheni to be representative of the way in which impacts became a matter of public dispute along the route of BTC? Yes and No. Certainly, a number of other locations on the pipeline route, many of which were in Georgia, attracted international interest. These specific disputes were unlikely be have been predicted in advance of their emergence. Along with the disputes in Dgvari and Sagrasheni they were both products of the operation of corporate responsibility, and also had an irreducible contingency (cf. Foucault 2002a: 225).

Nonetheless, it would be wrong to imagine that most disputes along the route of the pipeline involved the mediation of international financial institutions, technical experts and NGOs, as they did in Dgvari and Sagrasheni. Alongside these more or less formal disputes, which involved complaints to the IFC ombudsman, there were possibly numerous instances in Georgia, in particular, where villagers themselves took direct action, blocking construction work. In a town not far from Sagrasheni, villagers not involved in the dispute were said to have blocked construction work in order to stop the use of a stone crushing machine nearer to the town. Residents of Dgvari and Sakire also blocked construction work on occasions.[14] While some of these actions were reported in local media, there were only very brief accounts of such blockages in the English-language reports of Georgian environmentalists and the BTC SRAP review panel: 'in the latter part of 2003 and early 2004, BTC Co in Georgia has experienced numerous instances where the pipeline right of way has been blocked by protesting individuals or groups of villagers. BTC Co advises that since the SRAP visit in February 2004, the number of blockages has diminished' (BTC/SRAP 2004a: A-18.). Nonetheless, the problem remained and was considered sufficiently serious for top-level government officials to 'deliver public messages as to the illegality of blockages' in early 2004, and for BTC's social and resettlement advisory panel to 'consider placing blockage management under the responsibility of BTC Co's Field Security Officers rather than Community Liaison Officers' (BTC/SRAP 2004b: C-37). Later the panel noted that, while the BTC community investment programme had not prevented blockages and stoppages, 'it has been successful in making such actions more negotiable' (BTC/SRAP 2006a: C-17). On occasions, however, such confrontations escalated. Near to the city of Rustavi, apartment dwellers complained of 'intervention' by the Georgian police (BTC/SRAP 2004a: A-18).[15] It was reported that in January 2004 'around 150 residents of the Krtsanisi village, near Tbilisi, staged a protest rally outside the State Chancellery in the capital city, demanding to cease construction of the Baku-Tbilisi-Ceyhan oil pipeline via their village'; in June and August, riot police were said to have broken up anti-BTC rallies in the vicinity of the village (Green Alternative/CEE Bankwatch 2005: 36).[16] By June 2004 village protests had become commonplace, with BTC advisors reporting that there were a 'record 45 blockages' as construction work proceeded across the country (ibid.: 37, BTC/SRAP 2004b: C-21). These had become a 'widespread channel for communities to express discontent, for issues sometimes unrelated to the project' (ibid.: C-24). The events of Sagrasheni and Dgvari were mediated by environmental NGOs and scientists, and came to be discussed in meetings in Washington and London and reported extensively on the internet. By contrast the unruly disturbances and clashes that happened in Tbilisi and rural Georgia were not.[17] In this way, the measurement of impact played an important role in marking the limits of distinct forms of political engagement and public visibility (cf. Rancière 2001).

In his essay 'Opinion and the Crowd', the sociologist Gabriel Tarde argued that the crowd should be understood as quite distinct from the public (Tarde 2006 [1901], Laclau 2005). For Tarde, there was a difference between the irrational imitative behaviour of crowds and the formation of publics, the existence of which depends on the collective provision of information. 'The crowd', for Tarde, 'has something animal about it, for is it not a collection of psychic connections produced essentially by physical contacts?' (Tarde 2006 [1901]: 8). By contrast, publics are forged through the progressive emergence of currents of opinion: '[their] bond lies in the simultaneous conviction or passion and in their awareness of sharing at the same time an idea or a wish with a great number of men' (ibid.: 8–9). In Tarde's terms, those villagers who blocked construction along the route of the pipeline could be understood as crowds not publics. But while Tarde himself extolled the virtues of publics, the strength of crowds, in this case, was that their actions could have immediate financial consequences. The construction of a major infrastructure such as the BTC pipeline depends on the tight coordination of the movement of equipment and workers in time and space. As the villagers recognised, this is a process that can easily be disrupted at certain 'tactical points' through direct action (Watts 2005, Mitchell 2011). In this respect, the Georgian government's decision to bring a halt to the construction of the pipeline in Borjomi, a mountainous region at a critical moment in the summer months (see Chapter 2), perhaps showed that the state could imitate something of the actions of the crowd and its immediate physical presence.

Circulations of Impact

The BTC pipeline was expected to be visibly invisible. It was to be buried leaving 'little if any evidence' of its presence on the surface, minimising its impact on affected communities, while maximising its visibility through the production of information. But as this was so, any evidence of its visible environmental impact, whether it resulted from the movement of earth, trucks, air or pipes, could become politically significant. Here we have seen how signs of possible impacts – cracks in houses – came to be the focus for a series of grievances and public disputes. Understood as possible impacts, records of the existence of cracks circulated far beyond the route of the pipeline to the offices of scientists, environmentalists and officials in Tbilisi, London and Washington, DC. If affected communities were initially thought to be located in a well-defined corridor surrounding the pipeline, signs of impacts marked out a shifting and discontinuous space, one that could not be defined in advance.

In the following chapter I turn from the visibility, materiality and spatiality of environmental impacts to the physical materiality of the pipeline itself. As we shall see, the properties and behaviour of one specific feature of the pipeline, a coating material, also became – like the impacts discussed here – the object of intense dispute. However, while reports of the cracks of Dgvari and Sagrasheni circulated between Georgia and the offices of the World Bank, accounts of cracks in pipeline coating material lay at the heart of an inquiry by a group of British Members of Parliament in the House of Commons in London.

Chapter Seven
Material Politics

Thus far the material politics of the pipeline have been manifest in a series of disputes about the potential for landslides and the impact of construction work and construction traffic on, amongst other things, houses, roads, archaeological sites and biodiversity. In this chapter, however, I turn from these concerns to the materiality of the pipeline itself. This shift in the focus of analysis is linked to a spatial and political shift. As we have seen, the debates that erupted around the Borjomi route were particularly intense in Georgia, but also came to resonate in the US State Department; whereas the disputes that developed around environmental impacts such as vibration were addressed by the World Bank in Washington, as well as by Georgian environmentalists. By contrast, an intense dispute about the materiality of the pipeline arose in the House of Commons in London. In this setting, the material structure of the pipeline became, for a brief period, a political matter of a remarkably public kind.

This chapter is about this specific controversy and its contingency. Nonetheless, there are good reasons to use the study of metal structures such as pipelines to think about the relations between materiality and politics more generally. One reason is simply that there is something of a neglect of the politics of metals today, whether in terms of their extraction, manufacture, use or repair. If, as I will argue, the malleability of metals was once seen as an index of the transformative capacities of capitalism, today metals seem to have disappeared from view. We live, according to certain theorists, in a world marked by flows of knowledge and information, and in which

Material Politics: Disputes Along the Pipeline, First Edition. Andrew Barry.
© 2013 John Wiley & Sons, Ltd. Published 2013 by John Wiley & Sons, Ltd.

materials such as metals are no longer of great interest. Where once they lay at the heart of social theory, metals appear to have been relegated to the back stage. In what follows, however, I am not concerned with the malleability of metals; rather, I put forward a different thesis to the classical one: namely, that part of the interest of metal structures derives from their specific propensities and the contingency of their behaviour.

The chapter develops two arguments. One is that metals are not the hard, unchanging objects that they are often thought to be. Metal structures such as pipelines form part of dynamic assemblages in which the expertise of engineers, metallurgists and other material scientists have come to play a critical part. They have become informationally enriched, and part of the driving force for this enrichment comes from growing efforts both to regulate and enhance the properties of materials. In this respect, the BTC pipeline is an example of an informationally enriched or informed material structure, which includes a steel pipe, but includes other materials besides. The second argument focuses on a dispute that took place in the UK House of Commons, which revolved around the question of what was known by both BP and the UK government about the properties of a specific material component of the BTC pipeline. If political action often involves the staging of a particular issue as a matter of collective importance (Chapter 4), then non-human materials rather than human subjects were, in this instance, placed firmly centre stage (cf. Rancière 2004a, Bennett 2005). In reflecting on the dynamics of this particular dispute, I highlight the tension between the explicitly political staging of the issue in the House of Commons and the micro-politics of engineering expertise.

Metals

To begin, it would be a mistake to think that the study of metals is only a branch of physics. Indeed, from the point of view of the metallurgist or engineer, the properties of metals cannot simply be deduced from fundamental physical principles.[1] Alloys cannot be understood as combinations of pure substances, and the behaviour of metals in the conditions encountered in power stations or aircraft is quite different from any laboratory setting or simulation. Moreover, it would be a serious mistake to think that physics can simply be *applied* to the study of metals – or only if we take the term 'application' to involve the deviation of translation (Callon and Latour 1981). One of the preoccupations of the metallurgist (and I use the term very broadly to include all those concerned with the technical existence of metals and their relations to other substances)[2] is to be concerned with the specificity of the case, rather than account for the case in terms of general principles (cf. Berlant 2007). General principles are important, of course, but only in so far as they are not applied in any generalised way, and are acknowledged to be inadequate to the task in hand. The metallurgist *expects* that materials will be opaque, that the case will make a difference. In this way, the metallurgist is a good materialist, aware that materials will always, in some way, be resistant to external forces, and will generate their own effects (Stengers 1997). Not all may agree with this proposition, however: the socialist historian of science and crystallographer J.D. Bernal, writing in the early 1950s, reckoned that following the development of X-ray crystallography it would be possible for metallurgists and other scientists to begin to take 'rational control' over the internal structure of metals:

> The structural studies [following the development of X-crystallography] ... explained the primary, economically valuable properties of metals – their plasticity and hardening, the means by which metals can be forged, rolled and drawn – and made possible the beginning of a rational control of these processes. (Bernal 1969: 796)

While X-ray crystallography played a critical role in the development of molecular biology and solid state physics in the immediate post-war period, Bernal was over-enthusiastic about the possibility of turning metals into what we might call, following Foucault, docile objects. After all, X-ray crystallography is a technique that can only be used to determine internal structural features of carefully prepared specimens in a well-equipped laboratory. It cannot be applied directly to the study of metals in use, or in the field, where it is likely that they will be subject to variations of stress or temperature and the affects of chemical action.

In so far as metallurgy addresses the question of the relation between the transformation of metals and features of their external environment, it addresses a central problem for science and technology studies (STS). For STS was, of course, for a long time puzzled about the relation between 'external' (economic and social) forces and the shape of technologies. In this way, STS rediscovered a classical problem (Mackenzie 1996). In a remarkable passage, Marx formulated the relation between the historical development of capitalism, the division of labour in manufacture and the structure of metals precisely in terms of their shape: 'manufacture is characterised by the differentiation of the instruments of labour – a differentiation whereby tools of a given sort acquire fixed shapes, adapted to each particular application – and by the specialisation of these instruments, which allows full play to each special tool only in the hands of a specific kind of worker' (Marx 1973 [1867]: 460).

Contemporary metallurgy does not confine itself to external form and shape, however. Rather, one of the preoccupations of the metallurgist is with the question of how external forces and events become translated or absorbed at the level of molecular structure, and conversely how molecular structure is mediated in transformations of external form. As Roux and

Magnin argue, metallurgy is not so much the science of the microscopic *or* the macroscopic, but a mesoscopic field which mediates between scales and spaces, and between different forms and techniques of analysis (Roux and Magnin 2004: 11). The metallurgist is an expert who is capable of bringing different spaces and objects of analysis simultaneously into view, moving between observations of external and internal structure, between quantum physics, thermodynamics, corrosion chemistry, crystallography and management strategy, between idealised atomic models and phase diagrams and materials in use, between the human and non-human elements of assemblages (cf. Mackenzie 2002: 16).

From the point of view of contemporary metallurgy, metals are sites of transformation. Internally, they contain features such as grain boundaries, regular lattice structures, impurities, dislocations and catalytic sites, which provide the basis for both stability and rigidity *and* movement, elasticity and flow – for changes in their intensive and extensive properties. They are spaces within which minute changes occur routinely, and catastrophic failures may represent the crystallisation of a series of infinitesimal movements rather than the immediate impact of an external force (cf. Tarde 2001 [1890]). It is common enough in social theory to draw an opposition between the static, bounded or rigid and the fluid or mobile; indeed, for some social theorists, speaking of boundaries and rigidities at all is thought to be *passé*. But it would be wrong to oppose the solidity of metals to the liquidity of fluids, or boundedness to flow – as if these are dualistic and absolute states. Rather, it is a question of recognising that solidity may itself be the product of certain form of fluidity. After all, metals are extraordinarily fluid – full of local sources of transformation and instability; and in this sense they are actually more fluid than fluids. Indeed, Deleuze and Guattari took the insights of metallurgists to be an argument for vitalism: 'what metal and metallurgy bring to light is a life proper to matter, a vital state of matter as such, a material vitalism that doubtless exists everywhere but is ordinarily hidden or covered, rendered unrecognisable' (Deleuze and Guattari 1987: 411). The metallurgist is not just concerned with the shape or mould within which metals are formed, or with their malleability, but with what might be called the continuous modulation or variation of metals (ibid., see also Deleuze 1979).

So metals share certain properties with living materials, and they flow; it is just that they often flow more slowly and, from the point of view of the metallurgist, more profoundly and irreversibly than fluids. They can contain historical records of their own past, in a way that most fluids cannot. They have surfaces, but their surfaces are sites of transformation, such as corrosion and friction, as well as functioning as boundaries (Bowden and Tabor 2001). Metals' capacity to continue to exist over years and decades depends on fatigue and creep: the minute internal transformation of metals under fluctuating conditions of stress and temperature. So metals are quite unlike glass, which may shatter under the impact of an external force, or many fluids, which may simply move to another place under external pressure, adapting to the shape of the container in which they are placed. Metals have the capacity to render external energies into novel internal forms, 'modifying [themselves] through the invention of new internal structures' (Simondon 1992: 305). Metals are solid and hard and, for a period, they can endure without ever remaining the same.[3] Their stability as material forms is intimately associated with both their internal transformation, and their fragility (Roux and Magnin 2004).

But if metals have something of a metastable existence, passing slowly between states, they also come to exist in other forms generated through the work of metallurgists, and the demands of regulators. In Bensaude-Vincent and Stengers' account of the *History of Chemistry*, instead of merely imposing a *shape* on matter, chemists proffer a 'different notion of matter':

> Whether functional or structural, new materials are no longer intended to replace traditional materials. They are made to solve specific problems, and for this reason they embody a different notion of matter. Instead of imposing a shape on the mass of material, one develops an 'informed material' in the sense that the material structure becomes richer and richer in information. Accomplishing this requires detailed comprehension of the microscopic structure of materials, because it is in playing with these molecular, atomic and even subatomic structures that one can invent materials adapted to industrial demands and control the factors needed for their reproduction, whether they are new or traditional. (Bensaude-Vincent and Stengers 1996: 206)

The same observation applies to metallurgy. The product of the contemporary metallurgist's labour is not necessarily a new metal; it is likely to be the informational enrichment of materials, multiplying their forms of existence. Through the work of metallurgists, metals acquire multiple lives: in simulations, micrographs, X-ray crystallography, and samples taken from materials in use. In each of these settings, metals exist in different forms (more or less prepared, more or less purified, more or less isolated from other chemicals), which depend on particular informational-material practices of experiment and field research (cf. Mol 2002: 6, Barry 2005). Their informational enrichment is bound up, as Bensaude-Vincent and Stengers suggest, with the need to ensure that their molecular structure is adapted to specific industrial and regulatory demands.

Thus, metals not only have a lively existence, their existence increasingly depends on the manner and degree to which they have been informationally enriched. While metallurgy might not provide the basis for the level of control over the properties of metals envisaged by Bernal, it nonetheless plays a critical role in their management and government. Consider, for example, the importance of the tests and measurements that are routinely carried out on systems such as power stations, aircraft and oil platforms in order to ensure that their integrity and safety are not threatened by, for example, corrosion or fatigue. Such measurements are governmental acts: they are intended to manage the potentially unruly conduct of material assemblages, aligning them with broader economic and governmental objectives. Just as the regulation of drugs demands the multiplication of their forms of existence as informed materials, through *in vivo* and *in vitro* investigations and through clinical and pre-clinical trials, metals are also subject to a series of commercial and regulatory tests, the results of which may or may not be made public (cf. Barry 2005, McGoey 2007, Rosengarten 2009).

Metallurgy might be described as something of a social and political science, if we understand the notion of the social in the sense given to it by the sociologist, Gabriel Tarde. For Tarde it was possible to refer to atomic or molecular societies as well as human societies, and he argued that the same concepts could be used to refer to the societies described by the physical and life sciences as to those analysed by sociologists (Tarde 1999 [1893]). The metallurgist might follow Tarde in acknowledging that there is no discontinuity between the realm of the social and the natural, the human and the non-human, or between the informational and the material, the living and the non-living (Whatmore 2006, Barry and Thrift 2007, Thrift 2008). Whereas Bernal imagined that, through the use of techniques such as X-ray crystallography, metallurgy would make it possible to establish something like a socialist administration of metals, resulting in a direct alignment between the internal structure of metals and economic need, contemporary metallurgists pursue a more flexible approach. For metallurgy today assumes that there need be no correspondence between material and social and economic structures. Rather, the metallurgist multiplies the forms in which metals exist, while recognising that complete knowledge and control are impossible. Metallurgy is an interdisciplinary discipline concerned with the study of systems or assemblages, in which metals and other materials form only a part.[4]

Metals and metallurgy provide, then, a particularly good starting-point for thinking about the properties of materials. They clearly illustrate the principle of irreducibility: the behaviour of metals resists any reduction of their properties, whether to their external (social) environment, or to the fundamentals of physics. Metallurgists are mediators between forms of economic calculation, government regulation and the analysis of material properties and structures (cf. Osborne 2004). Moreover, metallurgy, like agricultural research, zoology, geology, engineering, anthropology and geography, is reliant on field research, an artisanal and itinerant mode of practice, and not just laboratory experimentation (Deleuze and Guattari 1987: 411, Schaffer 2003, Livingstone 2003). As a field science, metallurgy should in principle be attuned to the specificity of the case; it aspires to be attentive to the general problem of how to address the study of the particular. Where many physicists may be preoccupied by the problem of how to represent the particular in terms of the general, metallurgists are often confronted by a rather different question, namely: how is it possible to understand and manage the properties of objects that exhibit general problems – such as fracture, conduction, phase transition, creep or corrosion – but in specific ways, and in very different settings and locations?

Nonetheless, while there has been a longstanding interest in metals in social theory and philosophy, metals can rarely be found in isolation from other materials. They enter into material assemblages. I take this observation, in what follows, to what might appear initially to be a fairly straightforward technology, the BTC pipeline, which consists not only of steel but a range of other materials, which are themselves, moreover, in contact with aggregate, soil and water. The remainder of this chapter focuses on the remarkable significance of a very specific element of the material structure of the pipeline. This was a 'field joint coating material', SPC 2888, that covered the connections between different pipeline sections. This material is 'two-component liquid epoxy urethane', and it was claimed by its manufacturers to be fast curing, flexible, tough, and environmentally friendly, as well as capable of being applied in a single coat (House of Commons 2005b: ev26). While SPC 2888 is a non-metallic substance, the engineers involved in its development, testing and application were necessarily concerned with its relations with other materials, including the coating of the BTC pipe sections, the environment within which the pipe would be buried, and the range of temperatures to which the pipeline was likely to be subjected. They were engaged in a type of field research, and not just laboratory experimentation. Here I argue that the case of SPC 2888 raises two sets questions. The first concerns the relation between the properties and behaviour of materials and the organisation of economic and political life. How can particular materials, such as SPC 2888, come to be constituted as events of general significance to others (Barry 2002, Runciman 2006, Dewsbury 2007)? Secondly, how did political controversy come to focus on the unruly behaviour of materials rather than the behaviour of persons? And to what extent did the work of engineers, metallurgists and other material scientists come to have political agency?

Politics

In the last chapter we saw how the disputes that revolved around the houses of Sagrasheni and Dgvari came to the attention of the International Finance Corporation and the European Bank for Reconstruction and Development. By contrast, the dispute over the significance of SPC 2888 became particularly intense during the course of an enquiry by the UK parliament's House of Commons select committee on Trade and Industry into the activities of the UK government's Export Credit Guarantee Department (ECGD)

(House of Commons 2004, 2005a&b). The enquiry was focused on the operation of the Department's Business Principles, which were expected to govern the relation between the Department and the companies to whom it provided financial assistance. Yet although the enquiry had a very specific focus and examined the activities of particular and arguably minor government agency, these activities raised, according to critics, wider questions. Did the government exercise control over the behaviour of corporations in other countries, or does the government primarily act to facilitate corporations' activities? What is the character of relations between the government and multinational corporations?

While the remit of the select committee was to address the implementation and effectiveness of business principles by the ECGD it nonetheless came to focus on a particular example of the implementation of these principles. This was the financial support given by the ECGD, in conjunction with the International Finance Corporation (IFC) and the European Bank for Reconstruction of Development (EBRD), for the construction of the BTC pipeline.[5] As we have noted, the involvement of ECGD in the project was intended to reduce the financial risk to investors, but also helped to ensure that the UK government, in particular, would have a direct interest in the completion of the project. BP and other international oil companies were willing to be submitted to the greater scrutiny that the receipt of public finance would entail, in part because it would ensure that Western governments would have this interest.

However, even within this restricted focus on the financial support of the ECGD for the BTC pipeline, the select committee channelled its critical scrutiny still further. Prompted by the work of the Baku-Ceyhan Campaign, critical of the work of the ECGD, the committee devoted considerable time to the failure of a particular coating material, SPC 2888, used on joints between sections of the pipeline.[6] More precisely still, it was concerned with the very specific issue of what the ECGD knew about the procurement and use of this coating material in late 2003 during the period when the Department was considering whether to support the construction of the pipeline. Indeed, the case of the coating material was the primary issue related to the BTC pipeline discussed in the House of Commons, with the exception of a brief discussion of the case of the Kurdish nationalist activist, Ferhat Kaya, who was allegedly tortured or ill-treated in the police station in the Turkish town of Ardahan, near to the Georgian border, on account of his criticism of the BTC project (Baku-Ceyhan Campaign 2004a, House of Commons 2005b: ev 52).

The centrality of this particular coating material to the concerns of British Members of Parliament is moreover surprising when viewed in relation to debates elsewhere. For a time, during 2003–5, the pipeline acquired a remarkable political geography. In Washington, DC, in particular, BTC came to have a very different significance. As we have seen, the offices of the ombudsman of the International Finance Corporation on Pennsylvania Avenue, for example, investigated a series of specific alleged violations of Bank guidelines by the BTC company in Georgia, following representations made by the Georgian environmental NGO, Green Alternative (Chapter 6). Elsewhere in Washington, the State Department was forced to intervene following the decision of the Georgian government of Mikheil Saakashvili temporarily to halt construction of the pipeline in July 2004, and the issue was discussed in meetings between Saakashvili and Colin Powell and Donald Rumsfeld (Chapter 2). Moreover, the failure of the coating material described by the engineer had not led to any oil leak or effect on the environment. There was no specific accident to which anyone could point, although an investigative journalist who had researched the case described the coating material as an 'environmental time-bomb', while observing that campaigners were now demanding that BP 'live up to its ethical commitments' (Gillard 2004). Nor did the problem have any discernable impact on the complex geopolitical situation within which the pipeline was embedded. The failure of the coating material was not considered of particular importance by residents living near the pipeline route who were incensed by their failure to receive compensation that they had expected to receive due to the presence of oil industry construction work near to their homes (cf. Anderson 2006). In a meeting with Georgian workers and residents in the city of Rustavi nearby to the pipeline route, I was told that up to 50 km of pipeline had had to be re-laid.[7] But this was of little concern to the workers, who were angry about low pay, long working hours and poor food, and had been engaged in unofficial strike action in the same period (Chapter 8). However, the select committee did not consider the issue of working conditions and wages, even though it might reasonably have done so.

Why then should this committee, prompted by NGOs campaigning against the pipeline, take such particular interest in these cracks, and the specific issue of pipeline coating material, rather than the working conditions of Georgian pipeline workers or the partial exemption of BTC from the terms of Georgian labour law, for example? Why was the politics of a material considered more significant that the politics of class? (Gibson-Graham 2006). If politics, as Rancière suggests, involves making objects and problems visible, why were the failures of material objects rather than the working conditions of labourers rendered visible to the committee? (Rancière 2004b: 226). Why should defects in materials rather than defects in labour relations, pay and working conditions stand in for wider problems between business and the UK government or, more generally, between state and capital? Why, in this case, did the properties of materials come to have such political significance? (Barry 2001: 215)

An answer to these questions is complex. For if metallurgy is a form of field research which needs to address the specificity of the case, the same is true of field research concerned with the study of politics. An analysis of this event would involve consideration, for example, of the critical historical role of the Green movement in both Soviet and post-Soviet Georgian politics (Chapter 2).[8] It would involve examination of the particular timing of the failure of the coating material which occurred just before the decision of the ECGD to support the development of the pipeline. It would involve an analysis of how the question of the reputation of oil corporations has become a focus for both management and political action, leading to the increasing interest in corporate social responsibility (Chapter 4). Crucially, it would involve an assessment of the preoccupation with formal procedures of accountability and transparency in political and economic life (Chapter 3). In short, it would entail a series of investigations of the multiple political situations which both informed this particular controversy, and to which it contributed (Chapter 1).

The salience of the politics of materials rather than the politics of labour in the House of Commons also turns partly on the legitimacy of particular sources of evidence. After all, the problem of the coating material could not be denied, for everyone, including BP, accepted that it had happened. Long sections of pipeline had to be repaired as a consequence of the failure of the coating material. Once acknowledged, this could not be simply explained away by any suggestion that the failure of the coating material was conjured up by the opponents of the oil company or the government for political or

Figure 7.1   The Materiality of Construction. Photo taken by author

financial gain. Unlike the material demands by pipeline workers that they should work shorter hours and be paid at higher rates, claims concerning the existence of cracks could not so easily be accused of being self-interested or, indeed, even 'politically motivated'. In comparison to the protests of the workers, the materiality of cracks in the pipeline coating material was less clearly entangled in the complexities of Georgian politics in the aftermath of the Rose Revolution. And unlike the demands of Georgian workers, which were mediated by local lawyers and trade union representatives who in general did not speak English, the existence of cracks was mediated directly in London by NGOs and the media.[9]

In what follows, however, I focus more narrowly on the question of the presentation of evidence in the House of Commons. After all, the potential significance of evidence depends on the setting in which the evidence is presented and the audience to whom it is presented (Shapin and Schaffer 1985). In representing evidence of the failure of pipeline coating material in the House of Commons, NGOs sought to effect a radical translation in its significance. Evidence of the existence of material failure mattered in the House of Commons not primarily because it involved information about materials, and their local conditions of existence in use, but because of NGOs' sense of the materiality of this information in relation to the behaviour of the government and the multinational. The presentation of evidence in the House of Commons would have a quasi-legal effect, demonstrating the complicity of the multinational and its supporters in government in a public forum. In this setting, the particular was of little interest in terms of its particularity, but in so far as it provided the basis for a wider argument about the activities of the ECGD and its failure to act according to its principles. Moreover, the case of SPC 2888 could lead to inferences about the wider forms of complicity between corporate business and government. Did the government exercise control over the behaviour of corporations in other countries, or did the government primarily act to facilitate corporations' activities? Was the growing concern with the ethics and transparency of business in general, and the oil industry in particular (Chapters 3 and 4) reflected in the actions of government? Or, even more broadly, were the 'Business Principles' of the British government simply particular features of the operation of neo-liberalism or the 'neo-liberal state' (Harvey 2005).

But how was it possible to translate knowledge of the behaviour of materials in a specific locality, of no obvious significance to a group of parliamentarians, into information that was of material importance to the recommendations of a select committee? How could one translate a (technical) 'fact' about the failure of materials in the field into a (quasi-legal) 'fact', which would matter to the deliberations of a select committee and demonstrate the guilt of the government and the multinational (Latour 2004)?[10] Critical to the NGOs' case before the select committee was the testimony of a pipeline engineer concerning the period prior to the start of pipeline construction in 2003. This testimony was expected to acquire political agency once presented in the House of Commons.

In November 2003, at around the time of the Rose Revolution in Georgia that led to the end of the government of Eduard Shevardnadze, cracks in the material that covered the connections between separate sections of pipe emerged during the construction. The BTC company claimed that the cause of the fault was that the field joint coating covering the connections had been mis-applied as the temperature dropped in November, but that following further investigations and tests the problem had been rectified (House of Commons 2005b: ev26–27). Despite the previous existence of cracks in the coating material, the pipeline could be buried safely. The pipeline engineer, himself a consultant who had offered his services to BP, the major oil company involved in the BTC project, was incensed that the company had previously failed to think through the relations between their actions in selecting this particular coating material for the oil pipeline, and the behaviour of the pipeline in the field. The pipeline engineer explained to the parliamentarians:

> Oil and gas pipelines are not passive, inert items, they are *live, dynamic structures* that move due to ground movement and most importantly, pressure changes within the pipe ... The coating has to accommodate such movement. The operating temperature will fluctuate with pressure changes and should the pipeline be shut down for any time, the pipe temperature will drop down to the in-ground ambient – estimated by BP to be -5^0 to $+50^0$ C ... How will this affect the performance of the coating particularly at the PE/epoxy interface ...? This question has been discussed throughout the whole pipeline industry and I am yet to hear any individual say – 'it will be OK, the system is fully proven'. (House of Commons 2005b: ev60, emphasis added)

The pipeline engineer argued, furthermore, that the modified epoxy coating had been inadequately tested, that the specification for the coating was inadequate, documentation unsatisfactory, and that tried and tested alternatives were not properly considered. In short, using SPC 2888 involved a considerable and unnecessary risk. Another engineer who gave evidence observed that 'if you have something that does the job and these other systems have been extensively applied and [have] a working history why change and in particular to use this very important pipeline as a proving ground for an experiment with a new coating system' (House of Commons 2005b: ev85). For the NGOs and a journalist, the defects in SPC 2888 embodied defects in BP itself and its relations with ECGD and the lenders' group consultants who ECGD relied upon in their exercise of due diligence. These consultants, were, in effect, told by the lenders to rely on the integrity of BP in providing them with accurate information. This was a scandal: due diligence assumed that the company could be trusted even when there were those who were able to provide evidence to show why it should not be. The failure to investigate defects in coating material reflected wider defects in the activities of multinationals, banks and government, and their all-too-intimate relations: 'this statement [that the lenders' group did not want the problem examined further] provides an extraordinary insight into the approach taken by the lenders' group [including the ECGD]'(House of Commons 2005b: ev105).

But if the pipeline engineer's willingness to speak openly about his concerns with BP provided the opportunity for NGO critics to demonstrate the complicity of multinationals and government, his statement to the House of Commons points to a different kind of politics, and a different form of expertise, to that of the NGOs. If politics partly revolves around the question of how the particular is figured as an instance of interest to a collective, then the pipeline engineer's political concerns, and his understanding of the relation between the particular and the general, are quite distinct. For although the pipeline engineer spoke of cracks in coating materials, he viewed these as an index of a 'guinea pig engineering culture' that failed to attend to the liveliness of materials rather than as a sign of complicity between government and business in general. Nor does his evidence necessarily imply a link, for example, between the oil company's engineering culture and a series of other problems that I have discussed earlier. However, for NGO critics the failure of the coating material formed part of a series of other specific events and incidents that were all signs of the state of relations between government and the oil business. In this way, the case of SPC 2888 contributed to a wider abductive conclusion (Marriott and Minio-Paluello 2012: 175–177).

Although the pipeline engineer gave evidence, he also gave his evidence with passion and anger. He could not 'believe the crassness of the statements' in the report produced by the engineering consultants working for the lenders' group, which included ECGD. In speaking with such passion, he gave up the pretence that his evidence was, as the evidence of a scientist might be expected to be, dispassionate (Bennington 1994: 135). His anger conveys his sense of how badly particular elements – this steel, this soil, this coating material, the skills of these subcontractors, the winter climate of Georgia, and so on – had been assembled together. And he detailed the reasons why this occurred, with the specificity of this case that had so many surprising wider consequences. Engineering here stands as an example of an itinerant and artisanal practice which, potentially at least, addresses the impossibility of fully governing the behaviour of materials, taking proper notice of their differential resistance. The failure of materials is not surprising, because materials are not the dead, inert substances they are sometimes imagined to be. But the pipeline engineer was not disinterested in or unaffected by this particular case of material failure. In his account, the company had tried something out without having properly checked to see if it was going to work.[11] The pipeline engineer did not articulate any opposition to corporations in general. In my interpretation, his preoccupation was with the irreducibility of the properties of materials and a defence of the autonomy of his expertise in the behaviour of a material. His was a more-than-human politics (Whatmore 2006).

The significance of the pipeline engineer's testimony was judged in a public setting: the select committee (cf. Lynch 1998, Schaffer 2005). Within the UK parliament, select committees have a particular significance. As in the US Congress, a select committee is a group of politicians, selected from all parties, who interrogate the conduct of government and the development and implementation of legislation in public. A parliamentary committee is not a court of a law, for its recommendations do not carry the force of law. Nor is it a community of experts, for, although a select committee may seek expert advice and is likely to have its own expert advisor, it does not claim any expertise itself. Yet, like a court of law, a select committee is expected to function as a space where matters of fact can be established and judgements can be made on the basis of the evidence presented before it. Moreover, on account of the authority of parliament, it is able to request evidence and witnesses who may not be available otherwise and who, with exceptions, are required to give evidence in public. However, unlike the main chamber of the House of Commons, its final recommendations are expected, in general, to reflect the views of all of its members, and not just the views of the governing party or the statistical majority of the Members of Parliament (c.f. Waldron 1999a: 127). In this way, a select committee is potentially in the position to claim that its views are based on consideration of evidence and, at the same time, to be able to articulate, in principle, a non-party political agreement based on this consideration. Perhaps more than any other parliamentary institution, parliamentary committees claim to be able to act as 'modest (political) witnesses': ladies and gentlemen who confront evidence with disinterest (Shapin and Schaffer 1985, Latour and Weibel 2005) and yet who also represent the public interest. In effect they are thought to perform a function, regarded as essential in the institution of British parliamentary democracy, that it is possible to reach an agreement, not through consensus, and despite underlying disagreement, given the existence of an appropriate institutional mechanism and the prevalence of a certain form of ethical conduct in political life. At the same time, they were concerned to judge not just the veracity of the pipeline engineer's statement, but whether it was a matter of public concern. Should the failure in materials be an index of a wider failure in the relations between business and government? Should it even become an event that inaugurated a transformation in these relations?

Despite their exhaustive preoccupation with the circumstances surrounding the failure of SPC 2888, the parliamentarians ultimately were unconvinced about its wider significance. After all, their concern was with the behaviour of ECGD in relation to BP, and its adherence to its (ethical) 'Business Principles', not with the conduct of BP itself. 'It was not surprising', according to the select committee, 'that quality assurance problems occur during major construction projects such as the BTC pipeline. What matters is that those problems are identified and addressed' (House of Commons 2005a: 12). For the parliamentarians, the ECGD and the government had done all they could reasonably do to ensure that the problem of the pipeline coating was addressed: ECGD conduct had been 'proportionate and consistent with the Department's business principles' (ibid.: 13). They had done enough to investigate the properties of SPC 2888. As MPs they were not in a position to make a judgement about the behaviour of materials, only about the behaviour of government. And they based their judgement, in the manner of a court, not through commissioning a piece of independent field research on the situation in Georgia, or by talking to the workers of Rustavi or Gardabani, but on the basis of evidence presented before them (cf. Latour 2004: 101).

Nonetheless, there is no simple explanation of the parliamentarians' decision (cf. Law 2002: 143–162). To account for the decision one would need to consider the particular composition of the committee and its relations to government ministers, for example, and the level of trust of parliamentarians in BP in comparison to other UK companies. And one would need to examine the work of other pipeline engineers commissioned by both BP and the ECGD and the evidence they provided. The pipeline engineer's evidence was, after all, but one of a number of published and unpublished reports of the performance of the pipeline that circulated between Georgia and BP and government offices in Baku and London. There is moreover the question of whether an engineer, who expressed his views with such anger, was trusted by those who listened to his testimony. But in my reading, part of the reason why the evidence of the pipeline engineer was not thought to be a matter of wider concern is the way in which his intervention was framed by a situation that was too explicitly political. In effect, his evidence was placed in the midst of a conflict between corporations and governments, on the one hand, and their critics, on the other, which took place in the House of Commons. In this way, his concern with the specificity of materials, and the particular location and manner of their use, was understood too readily within this given political context. His micro-politics, which relied on his own understanding of the dynamic behaviour of materials, was over-interpreted in macro- or molar-political terms (cf. Deleuze and Guattari 1987: 216, Barry and Thrift 2007: 514). In this situation, the failure of materials and the pipeline engineer's evidence concerning this failure could not be made to matter beyond the confines of Parliament.

Itinerant Practices

Radical critics of capitalism have often developed their arguments either through an analysis of capitalism's systemic features and/or by making visible, through specific cases, the kinds of human misery, inequality and exploitation that are associated with capitalism's development. General analyses of capitalism's systemic properties have framed particular accounts, and specific examples have been taken as indices of systemic problems. In the case discussed here, the failure of material structures was taken by radical critics to be a sign of wider defects in the relations between government and business. This was a critical strategy grounded in a form of empiricism.

Yet if the behaviour of materials is sometimes taken to be an index of wider social relations, there is nothing intrinsically political about metals or other materials, or how they are shaped. If one common feature of political life is that specific issues or problems are made (for a time and in particular settings) into matters of collective or 'universal' significance (Zizek 2004: 70, Runciman 2006), and thereby become political, then there is no necessary reason why the behaviour or properties of specific materials should be considered a political matter. To be sure, forms of critical analysis can help them to become so, and yet such critical analysis can also interpret the political significance of materials in reductive ways (Mitchell 2002: 52). It is not inevitable that the behaviour of materials should be of interest to others, or that they should become the object of disagreement across a range of sites and settings within which political matters are addressed, whether in public or not. Materials acquire more-than-local political agency only occasionally, not in general. The political importance of materials arises therefore in particular circumstances and sites, such as those that emerged as we have seen in eastern Georgia in the winter of 2003–4. In this case, it depended on a series of contingencies, notably the coincidental timing of a stage in a decision-making process (whether or not to provide financial support for the construction of a pipeline), and a material event (the emergence of cracks in pipeline coating material). But it depended also on the behaviour of metals and liquid epoxy coating materials when applied in freezing conditions. It depended on the progressive formation of London as a centre of expertise and political debate concerning the question of corporate social responsibility in recent years. It depended on the preoccupation with formal processes of accountability, transparency and reputation in contemporary political and economic life, which made it possible for both an oil company and a government department to be accused of failing to be transparent, and for this to be considered potentially a matter of public political concern. And it depended on the existence of a parliamentary political assembly that, for a period, became interested in hearing evidence of the complicity between government and business.

In these circumstances, the analysis of knowledge controversies needs to attend irreducibly to a multiplicity of causes: to the timing and spacing of political life, and the moment and setting of politics, as well as the specificity of its techniques, institutions, forms of evidence and speech. But such an analysis should also address the ways in which the behaviour of metals and other materials plays a critical part in politics. Metals – and other materials – are not the inert objects they are sometimes imagined to be, merely shaped by social and economic forces; nor are they – as I have insisted in this book – the solid and stable foundation of social and political life. They are rather elements of lively and dynamic assemblages that may act in unanticipated ways, serving as the catalyst for controversies and thereby contributing to the transformation of political situations. Engineers and metallurgists are well aware of the difficulty of applying the general principles of physics and chemistry to particular cases, and of the need to recognise both the fragility of materials and the unpredictability of material processes. These lessons are also relevant to those concerned with the study of politics.

In the next chapter, I turn back from a focus on the material infrastructure of the BTC pipeline to its informational infrastructure: the archive. While this chapter has focused on the relation and the distinction between engineering and politics, the following chapter interrogates the constitution of the distinction between economy and politics. As we shall see, it introduces an array of additional material artefacts into the story of the politics of the pipeline, including beehives and trees.

Chapter Eight
Economy and the Archive

One of the powers of economics, Timothy Mitchell argues, is to help constitute 'the apparent border between the market and the nonmarket' (Mitchell 2007: 248, 2002). The archive of documentation produced by the company before and during the life of the pipeline was not the work of economists, but it was nonetheless intended, in part, to help forge a series of borders. In particular, the company differentiated its broadly economic activities from its wider ethical commitment to society, which took the form of a community investment programme. At the same time, the economy and environment of the region outside of the pipeline corridors was thinly sketched, and the politics of the states through which the pipeline ran was analysed only in the most general terms (BTC/RR 2003). In all of the archive there is little mention of, for example, the Georgian 'Rose Revolution' of November 2003 or the possibility of Russian intervention in South Ossetia. An analysis of what is conventionally understood to be politics is conspicuous by its absence. In *The Concept of the Political* Carl Schmitt noted that, 'in a very systematic fashion liberal thought evades or ignores state and politics and moves in a typical always recurring polarity of two heterogeneous spheres, namely ethics and economics, intellect and trade, education and property' (Schmitt 1996 [1932]: 70, Mouffe 2000: 99). The structure of the archive embodies this liberal ground as well as its core distinctions between economy, ethics and the state.

Yet if the archive has a particular form, it also continues to evolve. For the archive contains a series of documents, which record the performance of

Material Politics: Disputes Along the Pipeline, First Edition. Andrew Barry.
© 2013 John Wiley & Sons, Ltd. Published 2013 by John Wiley & Sons, Ltd.

the project, directing attention simultaneously towards the pipeline's recent past and its near future. They include the company's own Environmental and Social Report (which contained further reports on issues such as biorestoration and community investment), consultants' reports on Environmental and Social Compliance to export credit agencies and commercial lenders (BTC/IEC 2004: 5),[1] and the Social and Resettlement and Action Plan (SRAP) reviews 'to provide practical and troubleshooting advice to project management' (BTC/SRAP 2003a), carried out by experts in social development. In addition, there are reports dealing with the company's voluntary commitment to monitoring security and human rights, and reports by a group of well-known industrial and political figures, the Caspian Development Advisory Panel (CDAP), that reported directly to the senior management of BP. Finally, a few reports were produced by local NGOs in Azerbaijan and Georgia, which monitored a range of issues, including 'local content',[2] cultural heritage and reinstatement, during the period 2004–6.[3] Together these reports are manifestation of five 'layers' of 'monitoring, assurance and oversight', which were an intrinsic part of the transparency of the project (BTC/ESAP 2002a: 42, IFC 2006: 23). Four further 'internal' layers of monitoring were not made public.[4]

It would be a mistake to view this multi-layered complex and expanding body of reports as a smokescreen that covered up the 'real' operation of the pipeline. To be sure, their publication was no doubt intended and expected to have anti-political effects, enabling the company to demonstrate both that it had complied with a series of national and international guidelines and regulations, and that it could address emerging problems and risks. In principle, at least, the publication of this archive might be expected to have reduced the risk of unruly antagonism, anticipating, translating and diffusing the criticisms of NGOs and the potential opposition of affected communities, thereby rendering them as problems for management to address (cf. Hetherington 2011: 9). Yet, in practice, matters were more complex. For if the pipeline corridors were defined as much by the production of information as the occupation of land, then the borders of these corridors needed to be progressively refined, adjusted and defended through the production of more information. The story of the archive is, in part, in Stephen Collier's terms, a story of progressive 'accommodations and shifts' in the face of 'intransigent things and embedded norms' (Collier 2011: 242). At the same time, the transparency of the company's operations, embodied in this public archive, generated particular forms of dispute, which, in the short term could not be contained easily. The evolving archive registers some of these complexities.

In this chapter I consider three issues in turn. One concerns the relation and distinction between the economic activity of the company and its social and ethical commitments. Here I argue that these apparently distinct interventions could overlap, and become conflated, transformed or reworked (cf. Stoler 2009: 1). The documents that came to make up the archive described some of the constellation of processes and practice that came together in particular places (Massey 2005: 141), but also indicate some of the dynamics of their interaction and interference, and their emergent effects. Here I consider the interference between practices of compensation and community investment.

Secondly, the production of documents that came to form the archive could create intense feedback. The publication of information affected the world in a way that doubled back on the implementation of the project, generating noise, and producing instabilities and unanticipated consequences. Some of this feedback was highly localised and could take the form of village blockages and protests, or the planting of trees in anticipation of the possibility of compensation to come. But feedback could also be amplified elsewhere, far away from the pipeline, in Tbilisi and Baku, London and Washington, producing further loops, and more amplification. In what follows, I highlight the feedbacks between documents, the practices that they described and the world that these practices were expected to transform. The archive marks out distinctions between what should be included in the pipeline's corridors and what should be excluded. Indeed, statements contained in published documents were intended to be performative, as we have seen (cf. Mackenzie et al. 2007, Callon et al. 2007). They did not describe a society that pre-existed the pipeline's construction but were intended to help bring a society into existence. They defined and imagined the existence of entities such as 'project affected populations [PAPs]', 'communities', 'affected plots', and 'the construction corridor', which could subsequently become a focus for community investment or compensation (BTC/RAP 2002b). But if the company's statements were intended to create a society, the notion of perfomativity gives us a limited sense of the interaction between the production of documents and the society that they were intended to help form. Rather, this interaction can best be understood as complex, in the sense that the world was acutely sensitive to the production and publication of information about it, in ways that exceeded those that had been anticipated (Greco 2005: 24, Bell 2007: 116–117, Stengers 1997: 17). Here I read reports published over the lifetime of the project as evidence of the complexity and evolution of disputes to which the production and publication of earlier reports had themselves contributed.

Thirdly, as we have seen, the archive is marked both by an extraordinary level of detail, but also by systematic absences (Stoler 2009: 3). On the one hand, these documents are discreet about the activities of the state and other 'sensitive cultural, social and political issues' (IFC 2006: 5). On the other hand, while matters such as agricultural prices, vegetation (BTC/ OSR 2005), and the location of archaeological sites were mapped exhaustively, there is little account of the employment of construction workers or the complexities of existing forms of land ownership in the region, the significance of which became apparent as disputes over land ownership and compensation proliferated. If the route of the pipeline was the product of multiple histories and dynamics, some of these are clearly marked in the archive itself, while others are vaguely described, or simply absent. The archive renders particular aspects of the pipeline's construction hyper-visible but, in doing so, directs us towards the importance of institutions and processes that are only recorded in the margins of published documents. In effect, the archive creates a realm of what A.N. Whitehead terms the discernable – that which lies in the margins or the background and can only be indistinctly perceived (Whitehead 1920: 50).

Interference

After crossing the border with Azerbaijan the route of the pipeline loops round to the east of the industrial city of Rustavi before turning westwards to the south of Tbilisi, near to the village of Krtsanisi. It was here, in 2004, that a dispute developed over the possibility of a small grant from the BTC Community Investment Programme (CIP). The residents were all from one part of the village, a small 'village' of summer houses, provided by Eduard Shevardnadze for civil servants in the 1960s when he was Minister for Internal Affairs in Georgia (Suny 1994). However, the summer village was located in a barren area and had no running water. Moreover, because of military activity at a nearby base, the villagers felt compelled to let their cows graze on the pastures where the BTC pipeline was going to be constructed. Following the arrival of BTC in the area its residents hoped that a CIP grant would enable them to get access to running water. But there were two evident difficulties that needed to be addressed. Firstly, even if the residents were to receive a grant it would be insufficient to carry out the necessary work. Secondly, the summer 'village' was not officially recognised as a distinct village by the Georgian government and so was not considered eligible for a grant on its own, separately from other parts of the village. The second part of the village was occupied by Svans, who were displaced from their village in the Caucasus in the north-west of Georgia due to landslides, which are not uncommon in the mountainous Svaneti region. In the third part were ethnic Georgians, including refugees from Abkhazia displaced by the civil war that followed the break-up of the Soviet Union. The CIP had been intended to support 'sustainable development'[5] in the communities near the pipeline and foster good relations between the company and local people, thus benefiting the construction and subsequent operation of the pipeline.[6] Elsewhere in Georgia projects were already in process. But in the summer village, at least, it appeared impossible to achieve either objective.[7]

The village lay within two kilometres of the route of the pipeline. As such it could be the recipient of funding from the CIP, which was intended not to compensate communities, but 'empower local communities to resolve issues for themselves' (BTC 2006: 10-1). This programme represented a distinctly ethical form of economic intervention by the company. In Georgia, US$8 million was allocated to the programme in 2004–6, spread across the villages along the route of the pipeline during its construction phase. But, for the reasons given above, to describe the village as a community at this time would be to grant it too much unity and coherence (cf. Massey 2005). Indeed, it would be better not to view the village, in the context of this particular dispute, as a site of interference between different modalities of intervention that occurred over time. These included not just the company's interventions in the mid-2000s, but also the earlier interventions of the Georgian Republic during the Soviet period, which had established both the Svan part of the village and the summer village, as well as the interventions of the villagers themselves. After construction had finished in the area, BTC advisors noted that there were still 'some inter-communal conflicts' (BTC/SRAP 2005a: C-5).

What precisely were the company's interventions, according to published documents, and how were they differentiated? How were they framed or structured as distinct from each other? First, there was the commitment of the oil company to society, manifested in the CIP, and implemented by Mercy Corps in this region, and by Care International in western and southern Georgia. From the point of view of the NGOs and the oil company, villagers could potentially receive financial support, not because of the value of their labour or their land, nor because of the detrimental consequences of the construction of the pipeline on their immediate environment, but because of the commitment by the company and BP, its major partner, to the values of corporate responsibility. The CIP was a gift, for which communities were expected, according to company literature, to reciprocate by both developing good relations with the company, and by enacting the values of sustainability and self-empowerment (cf. Ssorin-Chaikov 2006: 357, Cruikshank 1999, Rajak 2011). The success of community investment depended explicitly on the agency of the community, which was expected to be both fostered and enrolled (Li 2007, Agrawal 2005). At this time, this gift was made available to all 'communities' along the pipeline route and communities themselves would determine how the money was spent, whether on infrastructural repairs to gas pipes or irrigation systems, the rehabilitation of the houses of culture that had been built in the Soviet period or repairs to village cemeteries. Later, in 2006, as pipeline construction ended, community investment came to be focused on villages that were able to write proposals that provided possibilities for their own economic self-development.[8] By the time of my return to the area in 2010 a flourishing dairy business had been established in the town of Marneuli to the south of Krtsanisi, enabling small producers to sell cheese to restaurants in Tbilisi (BTC/SCP/CDI 2011). By this time, CIP (now called the Community Development Initiative) was expected to make its own minor contribution to the formation of a market economy.

A further intervention was explicitly economic. As we have seen, the route of the pipeline had been mapped through a series of documents including the Environmental and Social Impact Assessment (ESIA) and Resettlement

Action Plan (RAP). The RAP detailed the ways those affected by the construction of the pipeline would be compensated. It specified the amounts of money that would be paid for land acquisition in particular regions, as well as the rates payable for losses in agricultural production. In total, US$8.6 million had been allocated for compensation payments in Georgia (BTC/RAP 2002f: 9-2) in order to mitigate losses 'in income and livelihood caused by the project' (ibid.: 1-1). Such compensation payments were governed by the terms of the World Bank's Operational Directive OD 4.30 and the IFC's guidelines which stipulated that 'economic displacement [that] results from an action that interrupts or eliminates people's access to productive assets without physically relocating the people themselves' should be recompensed (BTC/RAP 2002e: 1-9). In Çalişkan and Callon's terms, the company claimed to economise the length of the pipeline in multiple ways, both employing land and compensating farmers for any temporary losses they had incurred, purchasing labour and differentiating these largely individual financial transactions from the gift of community investment (Çalişkan and Callon 2009, see also Callon 1998a, Muniesa and Callon 2007). It sought to establish an economic zone within which land and labour could be purchased, governed by a distinct legal regime, differentiating this economic space from the wider economy of Georgia, Azerbaijan and Turkey.

Figure 8.1   Between geopolitics and corporate social responsibility, Krtsanisi, September 2010. Photo taken by author

Yet the location of the residents of the summer village pointed to the existence of a further modality of intervention in the vicinity, which is only briefly addressed in the archive. This was military and geopolitical. After all, there was also a Georgian military base in the vicinity. These were the sources of the unexploded munitions nearby which, together with the orders of the military commanders, discouraged the residents of the summer village from letting their cows graze near to the base. The base had twice previously been visited by US Defence Secretary Donald Rumsfeld; once in the aftermath of the September 11 attacks on the World Trade Center in 2001, and on a second occasion following the Rose Revolution in 2003. The Defence Secretary had inspected a detachment of US marines stationed at the base that provided support for the Georgian army under the $64 million Georgia Train and Equip Program (GTEP) (2002–4) and the subsequent Georgia Sustainment and Stability Operations Program (GSSOP). Reports document the role of US marines in training and in providing logistical support for the Georgian military which, according to the US government, would assist the Georgian government in its actions against suspected Al-Qaeda militants in the Pankisi Gorge in the north-east, near to the border with Chechnya (Global Security 2003). In this way, the area became part of the wider geographical scope of the 'war on terror' (Gregory 2004, Gregory and Pred 2007). Returning to the summer village after the pipeline had become operational, I was greeted by one of its residents, a retired Georgian interior ministry civil servant and former Red Army tank commander, who proudly wore combat trousers given to him by US marines, a visible reminder of this earlier intervention.

Thus, at this time, the summer village existed in the middle of three apparently distinct and powerful forms of intervention – geopolitical, social and economic – yet on account of its history (as a benefit for civil servants) and its precise location (near, but not directly adjacent, to the route of pipeline) it seemed unlikely to benefit from any of these interventions. As retired civil servants, the residents of the summer village did not own land in the narrow construction corridor of the pipeline, nor were they likely to be employed as workers by the pipeline construction company or its subcontractors. They were excluded from this process of economisation, thereby remaining dependent on the actions of the Georgian state, from which they expected little. The village remained trapped and immobile, for the time being, in an abandoned no-man's land (cf. Navaro-Yashin 2012).

Yet if the economic and social interventions of the oil company were conceived of as distinct from each other and distinct from the actions of national and foreign governments, these distinctions were not necessarily stable. Those involved in the CIP had sought to clarify the difference between community investment and compensation.[9] Certainly, the residents of the summer village were close to the route of the pipeline as it passed from the border with Azerbaijan before turning to the south of Tbilisi. And on the basis of their proximity they argued that they needed compensation, thereby confusing the company's distinction between, on the one hand, the ethical act of community investment (to support sustainable development and empowerment) and, on the other hand, land acquisition and compensation payments to farmers for losses in production during the period of pipeline construction. At the same time, while the company was not directly responsible for the presence of unexploded munitions, without the presence of the military the Georgian government could not claim to be able to maintain security along the route of the pipeline, and without US support, the pipeline would probably not have been routed through Georgia. In this way, the dispute between the villagers and the company over the possibility of community investment was also a dispute about the scope and dimensions of the political situation within which the village found itself.

The interference between economic, social and geopolitical interventions was not unique to this village. Nor was it only villagers who confused ethics, politics and the market.[10] The possibility of a relation between community investment and formal compensation payments was itself recognised by BTC's own advisory panel who had anticipated that the involvement of the CIP in conflict resolution might be necessary. In effect, the apparent distinction between the social and economic interventions of the company might need to be blurred as the political situation demanded: 'BTC Co.'s ongoing social monitoring program should be vigilant for any intra-village tensions between compensation recipients and non-recipients … Adjustments to the Community Investment Program to provide offsetting benefits may be required if such problems in communities are observed' (BTC/SRAP 2003b: A-15).

Yet if community investment was expected to foster good relations between BTC and villagers, the conduct of constructors could also interfere with the orientation of community investment towards empowerment and community investment. Whereas the CIP was intended to 'empower' villagers by expecting them to contribute to community investment projects, CIP workers noted that BTC construction contractors could undermine this ambition by doing work for communities – such as repairing irrigation channels or roads – in response to villagers' complaints or their attempts to block construction work, but without expecting any contribution from communities in return. Villagers could, by engaging in direct action, circumvent the need to become empowered through their involvement in the CIP, yet achieve similar ends and 'offsetting benefits'. While accounts of ongoing community investment projects figured in company publicity, entered into the public archive, and the CIP programme appears to have engendered few disputes in Georgia,[11] there is little published account of the informal actions of contractors, in their *ad hoc* 'in kind' response to villagers' concerns and protests. Nonetheless, BTC company advisors briefly confirmed the existence of this practice and noted the confusion that it produced with disapproval:

'The practice of some contractors negotiating "in-kind" compensation for use of land outside of the right of way is not in accordance with the RAP [Resettlement Action Plan] and should not be supported' (BTC/SRAP 2004a: A-11).

Compensation System

The dispute over the possibility of community investment for the summer village is unrecorded in the archive and not the focus of any reports by international NGOs. By comparison, highly public disputes proliferated along the route of the pipeline around the question of compensation, particularly in Georgia. At first sight this might be considered surprising. After all, one critical feature of the BTC project was the degree of its transparency in relation to the issues of compensation. In its documentation, the company had set out 'mechanisms for fair and transparent compensation for land acquired from private owners' (BTC/RAP 2002c: 1-10). By 2003 the company had published precise rates of compensation for losses for a vast range of agricultural products, from potatoes and honey to cherries and walnuts, appropriate to specific sections of the pipeline route, and had mapped out the existence of 'Project Affected Populations' and 'affected plots' to which compensation would be directed. In short, it had defined an economic space, which was intended to be performed. In practice, however, the publication of information produced complex feedbacks. The question of what was and what was not an 'affected' population or plot became a matter of dispute and concern and the company's claim that the process of compensation was transparent raised questions about what had not been made public. The transparency of compensation did not straightforwardly reduce the level of disagreement. Rather, it established a system that generated disputes of a particular form, which could be amplified and translated elsewhere, in London and Washington.

Across Azerbaijan and Turkey as well as Georgia there are numerous accounts of complaints about a lack of compensation for specific villages or individual villagers, published by both national and international NGOs. In Azerbaijan, it was alleged that individual land rights were sometimes changed during the land compensation process (OWRP 2004, Centre for Civic Initiatives et al. 2004). Meanwhile, international 'Fact-Finding Missions' to Turkey reported a number of specific instances of farmers who were unhappy about, amongst other things, the lack of compensation for the loss of trees and damage to village roads (Baku-Ceyhan Campaign 2004b: 31). In Georgia, Green Alternative along with the Georgian Young Lawyers, reported numerous cases of individuals who claimed not to have been properly compensated for the loss of land, trees or agricultural production. A later 'Fact-Finding Mission' to Georgia suggested, for example, that as many as twenty-six families in the village of Atskuri had not been adequately compensated for a variety of reasons. In particular, the contractor:

altered the pipeline route during the construction process. Yet compensation was awarded according to the original inventory: landowners originally designated as affected received compensation, while those actually affected did not. Where the landowner affected was the same both as planned and in actuality, compensation was not reassessed to represent different damage caused. This was confirmed by the Georgian Association for Protection of Landowners Rights (APLR). [...] Also in Atskuri, the pipeline corridor was widened from 44 m to 60–70 m in a number of places, as measured by the FFM [international NGO fact-finding mission]. No additional compensation was received. (Centre for Civic Initiatives et al. 2005a: 10)

In part, the transnational visibility of specific villages such as Atskuri, and others has parallels with the case of Haçibayram, discussed in Chapter 5. Cases of complaints about compensation made by individual villagers were cited by international NGOs as indicators that the project failed to meet international standards and guidelines more broadly. In the context of this politics, specific issues and individual cases could form the basis for abductive claims about the performance of the company and the international finance institutions (IFIs) in general. Yet, at the same time, the visibility of such cases cannot simply be understood in terms of their instrumental value to the NGOs' project. Indeed, the reports published by the company do not directly refute or accept the NGOs' claims. Rather, in diverse and uneven ways, they register some of the progressive adjustments of the company to a series of disputes. The particular form of the company's intervention, which sought to forge such a precise border between the transparency of the 'affected' corridor of the pipeline and its exterior, gave these numerous yet apparently minor disputes a particular dynamic, as well as, in some instances, a remarkable level of international visibility.

One broad set of reasons for the emergence of disputes can be discerned from the archive. Namely, there was a disjuncture between the precise divisions made in company documents and the complexity of the circumstances within which procedures set out in these documents had to be enacted. In principle, compensation payments were to be made to those who owned land within the 'construction corridor' itself, which was only 44 m wide in Georgia and 28 m wide in Turkey. Yet the narrowness and precision of this corridor did not correspond to other ways of dividing up the ownership and use of land, which earlier research commissioned by the company had not been sufficiently addressed. On the one hand, as Green Alternative and the Georgian Young Lawyers had observed, land use could be customary rather than based on formal ownership (CEE Bankwatch et al. 2003: 11–13), while some 'landowners [had] no clear understanding about the real location of owned land' (ibid.: 17). The archive records that in some instances landowners were now absent. For example, the rural population of Pontic Greeks in Georgia had declined substantially since the break-up of the Soviet Union, as many migrated to Greece or the north-east coast of the Black Sea (BTC/SRAP 2004b: C-40, BTC/SRAP 2006a: C-46). In these circumstances, the BTC company made use of a provision in Georgian law that allowed a landowner a right of way between two parcels of their own land, as a way of circumventing the problem (BTC/SRAP 2003b: C-12). On the other hand, if the area of the land that was to be purchased or leased was so limited then land ownership records needed to be very precise if compensation was to be determined fairly (IFC 2006). However, precise or accurate records did not necessarily exist (BTC/SRAP 2003b: C-5, BTC/SRAP 2004b: C-42), and/or it was disputed whether they were correct or not (BTC/SRAP 2003b: C-6).

The economic intervention of the company came in the wake of an earlier programme of land privatisation in Georgia and Azerbaijan, yet this had created its own legacy (cf. Mitchell 2005, Verdery 2003, Yalçin-Heckmann 2010).[12] Some of those on the list of landowners were said to have no connection with the land, and according to NGOs 'people have grave doubts that the inventory process was carried out fairly' (CEE Bankwatch 2003: 17). BTC's external monitors noted that there had been reports of District Prosecutors 'investigating some possible cases of fraudulent land registration' although they had not been able substantiate this (BTC/SRAP 2003b: C-20, see also Green Alternative 2005). In the Armenian village of Tabatskuri in central Georgia, 'the initial land registration done at the time land was privatised in Georgia was cancelled following an irregularity in the allocation process (plots were allocated to individuals whereas Georgian law provides they should have been allocated to households)' (BTC/SRAP 2004b: C-40). As a result, the company's initial offers of compensation were cancelled, and serious tension developed in the village, resulting in a blockage of construction work (ibid.: C-40–41). In the village of Moliti, in the Borjomi area, land registration had simply not taken place (ibid.: C-42). As Julia Elychar reminds us 'to intervene in any given social reality, an institutional power first has to map it' (Elychar 2005: 74). A map had been made, but it did not contain many of the significant features of the reality that was intended to be the object of intervention, leading to disputes between the company and landowners, as well as villagers who had customary use of communal land. According to the BTC company by late 2004 there 'had been 87 cases of disputed ownership taken through the formal grievance mechanism ... in addition APLR has collected documentation concerning a further 12 cases where they believe that information provided by the state may have been incorrect' (BTC/ESR 2004a: 7-4).

The difficulty of enacting the construction corridor as it had been represented in earlier documents was not only a function of the experience and political history of land ownership, which was not recorded, or the failure to anticipate this complexity in advance, but also the commercial logic and technological imperatives that drove the progress of construction. In some instances the route could shift in order to bypass unfavourable terrain leading to the possibility of payments for two routes, or one which was different from the one that had been expected,[13] leading to allegations that compensation had be given for the planned route but not the actual route (CEE Bankwatch et al. 2004: 16). At the same time, construction work was not contained within the narrow corridors mapped out in earlier plans, due to the demand for soil stockpiles, discharge ponds, and temporary construction roads, generating requests from contractors for 'additional land' (BTC/SRAP 2004a: A-10, BTC/IEC 2004: 48). The physical presence of large quantities of soil, water and waste, as well as the movement of lorries and construction equipment, progressively reconfigured the borders of the original 'construction corridor' in unpredictable ways. In these circumstances, the material borders of the corridor were not stable or necessarily well defined. By the summer of 2004, no less than 583 separate parcels of land had been impacted in this way, yet 'only 15% of these plots had been compensated by the contractor' (BTC/SRAP 2004b: C-11). The movement of lorries, pipes and waste was one key source of dispute as we have seen in Chapter 6. Compensation for the acquisition of 'additional land' was another: 'in Georgia, disputes with local communities have arisen where the construction contractor has constructed temporary access roads without adequately confirming land ownership or consulting with adjacent villages' (BTC/SRAP 2004a: A-11, Baku-Ceyhan Campaign 2005: 12).

But the transparency of the project could foster other dynamics. In particular, company documents could be read by villagers in terms of what they said, or they could be read in terms of what they did not say, or were thought to conceal, whether deliberately or not. The documentary film by the Georgian actor and director, Nino Kirtadze, illustrates this point strikingly. The film, which was largely shot in the villages of Tadzrisi and Sakire close to Dgvari in the Borjomi region in the summer of 2004, focuses on the negotiations and disputes between villagers and BTC over land compensation. The film cuts between scenes from the village and inside the offices of BTC in Tbilisi, tracing the translation of words and images from the oil company into the village and the actions of villagers back into the company offices. But, at the same time, the evolving political situation in Tadzrisi and Sakire is structured around a conflict within the villages. The disagreement is between a village leader and a school teacher over BTC's published statements about compensation and environmental impacts. The village leader suggests that the documents should be taken at face value as accounts of what BTC genuinely intended to do and the rates that it intended to pay; he states 'Ed Johnson [the head of BP in Georgia] is a capitalist, but he's not a fool' (Kirtadze 2005). The teacher, on the other hand, together with the other villagers, believes that the company is concealing something about the safety of the pipeline, which it refused to reveal; as one villager says, 'the pipeline will destroy us'. A third reading, which progressively emerges during the course of the documentary, is that it is the Georgian government that has misled BTC and this fact is not addressed in the documents published by the company. In short, the publication of documents points to the existence of processes about which little was published. In a concluding scene of the film, Johnson is applauded by the villagers when he accepts the villagers' contention that he needs to be concerned with the accuracy of the information provided to BTC by the Georgian government or, in other words, with a question — of the relations between the Georgian government and BTC — which had not explicitly been addressed in the public archive.[14] As the process of land acquisition came to an end, the IFC lamented the lack of attention to the complexities of land ownership earlier (IFC 2006). But this neglect was not surprising, given the discretion exercised by both the oil company and the IFIs about the organisation and everyday practice of the state. By contrast, the situated political knowledge of the villagers appeared more attuned to the need to think across the borders between the realms of economy and state politics, and address the question of the relation between the pipeline corridors and the territory through which they passed (cf. Corbridge et al. 2005: 190).

In the early 2000s, there was little market for land in rural Georgia. How then could the price of land paid by the company be determined? According to a number of documents contained in the archive, prices paid for land were high relative to rural income levels during this period. This view had been reflected in the company's Resettlement Action Plan although, in this document, it was thought to be part of a solution to potential problems rather than a source of problems to come: 'the SLRF [state land replacement fee] that will be paid for the affected land is close to 96 per cent of total annual household expenditures, such a *high premium* may help alleviate some of the potential problems with land acquisition' (BTC/RAP 2002b: 4-14, emphasis added, see also BTC/RAP 2002c: 5-12, BTC/SRAP 2003b). At this time the land price determined by BTC generally ranged between 20,400 to 68,002 GEL (US$10,000–34,000) per hectare depending upon the location and quality of the land, while the average monthly incomes of families near to the route of the pipeline were said to be as little as 344 GEL (US$172) per month (BTC/RAP 2002b: 4-11), although Green Alternative and the Georgian Young Lawyers observed that some landowners were not happy with both the lack of consultation about the land acquisition and the prices they had been given (CEE Bankwatch et al. 2003: 19). BTC's advisors on land acquisition and resettlement reckoned nonetheless that the price was relatively high: 'as a consequence of BTC Co.'s conservative strategy to purchase ownership rights to the construction corridor, and adoption of the State Land Replacement Fee as the basis for determining land compensation, project offered prices are well above those realized in the small number of recent land market transactions, or valuations based on Net Present Value' (BTC/SRAP 2003a: A-14). The assets of villagers could generally not be priced by the market, but were based on the charge payable to the state when land was changed from agricultural to industrial use. In effect, the Georgian state had not been external to the operation of the economy of the pipeline, but played a vital role in determining the value of land through which the pipeline passed, from which some landowners would profit.

Moreover, according to the archive, the relatively high level of land compensation payments was a second source of disputes. Some villagers complained about how much other individuals and other towns and villages were receiving: 'Bakuriani and Akalsikhe have received compensation even though they have no relation to the pipeline!'. The SRAP panel had warned that relatively high prices could give rise to 'intra-community tension between beneficiaries' (BTC/SRAP 2003a: A-31), noting that 'an adjunct to the generous compensation rates received by those with project affected land (referred to by villagers as the "lucky ones"), was the disappointment of those who missed out' (BTC/SRAP 2003b: C-17). This resulted in what they termed the paradox, which they later saw materialise in Tabatskuri, that 'people were struggling to be affected by the pipeline, rather than struggling not to be affected' (BTC/SRAP 2004b: C-7). However, some of those who received compensation might be concerned about why they were receiving so much. A woman, living in an area in which household incomes had been reported to be 277 GEL per month (BTC/RAP 2002b: 4-11), having been offered 5000 GEL (US$2,500) for a parcel of her land, asked me anxiously whether the pipeline would cause radiation, echoing the concerns of the residents of the summer village near to Tbilisi. The implication was that if the compensation was so high, surely it meant that the pipeline had to be dangerous or damaging. How can these anxieties be understood? Was her concern with radiation linked to a memory of Chernobyl or, as one Georgian BTC employee suggested, to the signs advertising the dangers of radiation that had been placed next to the pipeline during its construction on the occasions when X-rays had been used to detect defects in welds (Kirtadze 2005)?

Paradoxically, as the payments were considered by some to be large in comparison to prevailing income levels and market prices, they could also be considered simultaneously to be quite small. They could be taken to be compensation for a risk, from radioactivity, terrorism or natural hazards, which had yet to be properly acknowledged. After all, the opposition of Shevardnadze's Minister of Environment to the Borjomi route pointed to the existence of unknown threats (Chapter 2, Kirtadze 2005). Unsurprisingly, according to the company's own survey, over half of the 'project affected population' in the Borjomi region had been concerned that the project would increase 'the hazard of natural disasters' (BTC/RAP 2002b: 4-25). If the pipeline was potentially dangerous shouldn't the level of compensation actually have been higher? Moreover, as Marilyn Strathern notes, high expectations for compensation may not be an index of the loss suffered at all, but of a sense of the resources and energies of the company from which compensation was demanded (Strathern 1999: 189). With some justification a villager could both say that his land was worthless in the sense that it had no market value but, at the same time, argue that the level of compensation should have been much higher given the profits that the company would make through their ownership of the land. Demands could also be based on a sense of injustice (Gilmartin 2009), which could not be calculated, contributing to a sense that the compensation price had little relation to the loss of capital or income suffered by those who were compensated. During the development of the BTC project, few seem to have argued that compensation might have generated less conflict if the payments had been distributed more evenly, and not just directed at those households who owned land along the route of the pipeline itself. Later, however, one assessment of the process of land compensation noted, without further explanation, that 'the high price of land in Georgia has not necessarily led to the smooth implementation of the RAP' (BTC/SRAP 2005a: C-8). According to a different system of calculation based on an account of the value of the land to the company, the price could also be viewed as too low.

A third source of dispute derived directly from the operation of transparency. On the one hand, the standardization of prices could lead to the complaint from farmers that they, rather than the company, knew exactly how much income could be generated from a specific piece of land (OWRP 2004). But on the other hand, because the levels of payments made to compensate for losses in crop production were transparent, it was possible for landowners to calculate the potential for future compensation, which could be substantial, running up to tens of thousands of lari. Informants observed that in some locations, trees or flowers were planted near to the pipeline route in anticipation of compensation to come. In principle, transparency opened up the possibility of generating income by making use of the information provided. A landowner, for example, might claim to have planted roses along the pipeline route shortly before compensation was determined, and demand compensation to cover the loss of production over several years. In one village residents alleged that local officials had planted walnut trees, which were associated with particularly high levels of compensation. The BP CEO, John Browne, observed later in his memoirs that an 'unusual claim ... involved a grove of walnut trees which appeared overnight somewhere in Georgia' (Browne 2010: 172).[15]

In these circumstances, the question of whether trees had existed but had been destroyed by construction work, or had been planted simply in order to be destroyed subsequently, or had never existed in the first place, became fiercely contested: 'how do you know that my neighbour has not planted trees himself?' a villager asks a BTC employee in Kirtadze's film (Kirtadze 2005). The villager's question raises a series of further questions. What source of authoritative knowledge existed that could determine whether a claim for compensation was valid? How could it be shown whether a crop had been produced along the route of the pipeline, or whether a tree had once existed but now did not? How could a villager demonstrate that a tree had not been moved if some people were suspected by BTC to have uprooted trees and put them alongside the route of the pipeline? The presence or ownership of trees, which might be thought to have definite existence and location, could be difficult to determine, given the discrepancies between various records, including satellite images.[16] In these circumstances, the question of whether a given tree had grown for some time in the same place, or was a so-called 'magic tree' that had suddenly materialised,[17] could become a matter of dispute between the company and individual landowners. Following the completion of construction, BTC advisors noted that 'there are 13 refusers in Gardabani ... where the problem relates to issues of trees and rose bushes planted immediately prior to land entry at unsupportable densities' (BTC/SRAP 2005a: C-5). On occasions, international Fact-Finding Missions became interested in these disputes over the existence of trees and, in some instances, they were also mediated by officials from the World Bank or EBRD.

The same problem arose in relation to the location of beehives. According to World Bank guidelines beekeepers should receive compensation for losses in honey production due to noise and vibration if their beehives were located within 300 metres of a construction site. However, this led to a problem, and a series of disputes along the route of the pipeline through Georgia (BTC/SRAP 2005b: C-16, Chapter 6). Some beekeepers complained that they were not receiving the compensation to which they were entitled, while the company had to determine whether a beehive had always been near to the route of the pipeline or had been moved closer to the route in order that the beekeeper could benefit from compensation. To resolve this problem, BTC hired a retired scientist from the Georgian Ministry of Agriculture, who had once written a thesis on how to cultivate high-quality queen bees. He explained that the question of the relation between construction work and bees is a complex one. After all, bees are fragile and sensitive and they respond to many different changes in their environment including sunlight, heat, dust and vibration. This makes it difficult to determine specific causes for changes in honey production. Nonetheless, he was certain that the kind of noise and vibration associated with the construction of the pipeline, resulting from blasting and heavy construction traffic, would affect bees. He explained that he had previously conducted a field experiment in western Georgia when there was blasting and rock-breaking approximately 500 metres away from his research site. Although at this distance there had been little effect on honey production, he thought that the World Bank's guideline was reasonable enough.

In arriving at an assessment of the strength of individual beekeepers' claims, the bee expert relied on his own long experience of research and beekeeping as well as direct observation. In visiting a beekeeper he first interviewed him, asking how many bees had died during the construction and whether he had a logbook of how many bees he had in honey production, although usually he didn't. After the interview he asked the beekeeper for evidence of everything he had told him, and inspected individual hives. If the hives had been occupied there should be a bare patch of grass under the hive. But often hives had been moved closer to the pipeline in order to be found inside the World Bank's 300-metre corridor and the productivity of individual hives is often exaggerated. In total, he estimated that two-thirds of beekeepers had not told the truth, either about the location of their hives or their productivity. The expert in beekeeping estimated that each beehive might produce 70–100 Georgian lari (US$35–50) worth of honey per year and that most beekeepers have less than 40 hives. Moreover, even when the bee expert believed the beekeeper, the answer to the question of whether a loss in production was caused by the pipeline was complicated. After all, the hives and the pipeline do not together form an isolated system. Aspects of the environment of a hive were clearly important, but nonetheless difficult to assess. In 2004 there was an early spring, followed by a frost in April that had killed off flowers, resulting in low honey yields throughout the country.[18]

One beekeeper in particular, from the village of Bashkovi, took his case to the World Bank ombudsman, and was also visited by an NGO fact-finding mission (IFC/CAO 2004a: 4, Baku-Ceyhan Campaign 2004a: 24). While the ombudsman accepted BTC's claim that the complainant's hives were further than 300 metres away[19] from the Right of Way and therefore normally ineligible for compensation, there were special circumstances. For while it was not possible to claim that the bees could not have been affected by pipeline construction according to this rule, the beekeeper himself was as an internally displaced person (IDP) and needed to be treated as such. His status as a landless peasant warranted his treatment as a member of a 'vulnerable group subject to particular commitments under the Resettlement Action Plan', and his complaint was 'specific enough not to create a precedent' (IFC/CAO 2004a: 4). In these circumstances, as the complainant was a vulnerable beekeeper and not just a keeper of vulnerable bees the ombudsman welcomed 'BTC co's willingness to send the beekeeping expert to review impacts' (ibid.). What might be assumed to be narrowly a question of the costs of the impact of pipeline construction on honey production became, in this way, entangled with the question of the formal responsibility of the company and the IFIs towards the needs of 'vulnerable groups'.[20]

The World Bank ombudsman came to visit the beekeeper of Bashkovi, along with the residents of Sagrasheni and Dgvari and a number of other villages along the pipeline route (Chapter 6). Other experts from the IFIs, the oil company and their consultants visited and revisited a series of other villages, where other complaints had been made. The residents of Atskuri, for example, pursued their claims with the European Bank for Reconstruction and Development (EBRD), leading to the mediation of the EBRD Independent Recourse Review Mechanism (EBRD/IRM 2009, CEE Bankwatch 2008). Nonetheless, by 2008 the bulk of these disputes had formally been resolved, even if the reasons why the disputes had happened in the first place remain uncertain and disputed. The IFC reported that one of the 'lessons learned' from the project was that 'sponsors need to be prepared for the possibility [of significant complaints] and be able to source additional skilled resources to manage the process' (IFC 2006: 30). Certainly, the company had underestimated the need for such resources, initially deploying insufficient Georgian community liaison officers along the route of the pipeline and apparently failing to anticipate the number of disputes that their intervention would generate. But while the lesson drawn by the IFC is reasonable enough, it failed to consider how 'significant complaints' were themselves bound up with the way the pipeline corridors were both determined and rendered transparent.

Accounting for Labour

If the practice of compensation became, for a period, a matter of transnational as well as local dispute, what about the economic activity of labour? In what way did this, or did this not, become a political matter? In what way was the value of labour made public and contested? In this section I contrast the transnational visibility of disputes over land compensation with the local visibility of disputes over the value of labour.

In their re-reading of *Capital*, J.K. Gibson-Graham et al. stress the critical importance of accounting to Marx's analysis of politics of labour: 'like all systems of accounting', they note, 'Marx's language of class highlights certain processes and obscures others, potentiates certain identities and suppresses others, and has the capacity to energize certain kinds of activities and actors while leaving others unmoved. As a movable boundary, the distinction between necessary and surplus labour has made exploitation a visible and tangible object of discourse and politics' (Gibson-Graham et al. 2001: 8–9). Marx's thought, they argue, should not be understood as a more or less sophisticated or one-dimensional analysis of capitalism. Rather, 'his work was a political intervention, not only into specific political contexts but into the very meaning of politics and the range of social possibilities that politics avails' (ibid.: 9).

But if Marx's analysis of surplus labour and value in *Capital* can be understood both as a form of critical accounting and a political intervention, the empirical basis of his account derived, in part, from a further form of accounting: the evidence of factory inspectors. These inspectors, noted Marx, 'provide regular and official statistics of the voracious appetite of the capitalists for surplus labour' (Marx 1976 [1867]: 349). This appetite, he argued, extended into the minutiae of the working day, such as, in particular, the length of breaks. At the same time, factory-owners sought to avoid the gaze of the inspector. Indeed, the fact that factory-owners routinely broke the terms of the Factory Act, regarding matters such as maximum working hours, was clear enough. 'It is evident', Marx argued, quoting a report from a factory inspector, 'in this atmosphere the formation of surplus value by surplus labour is no secret, "if you allow (as I was informed by a highly respectable master) to work only ten minutes in the day overtime, you put one thousand in my pocket".' 'Moments', Marx observed, 'are the elements of profit' (ibid.: 352).

Yet while Marx, writing in the 1860s, was able to draw on the extensive and detailed reports of government factory inspectors, accounts of the working conditions and pay of construction workers in the international oil industry are much more limited (cf. Woolfson et al. 1996). For whereas both the oil company and international NGOs published extensive accounts of the social and environmental impact of the pipeline, these organisations showed little interest in inspecting the operation of construction work in a manner comparable to the factory inspectors. Outside of Georgia, Azerbaijan and eastern Turkey, the labour of local workers – primarily employed in unskilled or semi-skilled jobs for subcontractors – remained largely invisible. The wages and working conditions of local labour as well as workers from India, North Africa, Europe and North America were not discussed extensively by BTC and the international financial institutions in published documents, nor were they a central concern to those international NGOs who were critical of the project (CEE Bankwatch et al. 2004: 28).

Given the international interest in the pipeline during this period, this observation may seem surprising. After all, the question of pay and conditions amongst pipeline workers had became apparent in Georgia following a series of unofficial strikes during the winter of 2003–4, during the period in which cracks emerged in the SPC 2888 pipeline coating material in the eastern part of the country (see Chapter 7). Indeed, some of the workers from the eastern Georgian city of Rustavi, who were involved in industrial action, re-laid sections of pipes. Yet while the occurrence of cracks in the coating material acquired political significance in the House of Commons, the action of Georgian workers did not. The strikes themselves were animated, in part, by what Marx had termed 'moments' in the labour process. Workers complained about bad food, camp conditions, working hours, inadequate compensation for long journey times, low pay, short-term contracts, overtime payments, and requirements to work on election days.[21] Moreover, they claimed that whereas skilled workers, such as pipe welders, translators and crane operators, had been paid US$2.5 per hour in the previous year, they were subsequently paid a maximum of 2 GEL ($1) per hour. Semi-skilled workers, such as mechanics and scaffolders earned 1.5 GEL per hour while the lowest paid workers ('flagmen') received 0.75 GEL per hour.[22] One Georgian media source resented the fact that Georgian workers were paid less than Indian, Colombian and North African workers despite having what they claimed to be the same jobs. In practice, migrant workers were likely to be more skilled and more experienced.[23] In order to earn a wage of 600–700 GEL ($300–350) per month workers claimed that they would have to work 14 hours per day, including weekends and holidays.[24] Occurring in the period following the 'Rose Revolution' one informant regarded the strikes as politically influenced and the workers' claims as false. Nonetheless, similar issues, concerning working hours, food and differences in pay rates between local workers and foreign nationals were raised in Azerbaijan.

Although the initial series of unofficial strikes in Georgia was unsuccessful in leading to any significant improvement in pay, it directed the attention of the Georgian Trade Unions to the problem – as well as the NGOs Green Alternative and the Georgian Young Lawyers Association – who were convinced that the working conditions broke the terms of the Georgian labour code (Green Alternative 2003). The 'Committee Protecting Oil Worker's Rights' in Azerbaijan made the same argument.[25] In this respect, the NGOs and trade unions may have been right. But they had not recognised, at least at this time, that the terms of the Georgian and Azerbaijan labour code had already been partially superseded under the terms of the Host Government Agreements (HGA) and did not apply to the construction of the pipeline in important respects. The relevant clause of the HGA with Georgia stated:

> Subject to requirement that no Project Participant shall be required to follow any employment practices or standards that (i) *exceed those international labor standards or practices which are customary in international transportation projects* or (ii) are contrary to the goal of promoting an efficient and motivated workforce, all employment programmes and practices applicable to citizens of the State on the Project of the Territory, including hours of work, leave, remuneration, fringe benefits and occupational and safety standards, shall not be less beneficial than is provided by the Georgian labor legislation generally applicable to its citizenry. (Host Government Agreement 2000b, 18.2, emphasis added)[26]

In other words, contractors could ignore Georgian labour legislation if it was 'contrary to the goal of promoting an efficient and motivated workforce' or if legislation went beyond customary industry standards. One justification given for the formation of this neo-liberal space was that the

HGA was thought to anticipate the future of Georgian labour law and act as a benchmark for domestic legislation to come. The agreement was to be a 'forward looking' one. In other words, the existing regulatory regime had to be understood as a vestige of a different system that was informed by a different and out-dated set of values. This vision of the role of the HGA echoes the analysis, widely articulated in the 1990s, that the former socialist countries were in a process of 'transition' from a centrally planned socialist economic system to a market economy (e.g. Sachs 1996, Blanchard 1997). The notion of transition has been widely criticised by anthropologists in both positing a teleological movement from socialism to (free market) capitalism and giving 'capitalism' and 'socialism' identities as unified economic systems that they never had (Gibson-Graham 1996, Mitchell 2000, Humphrey 2002, Dunn 2005, Collier 2011).

The HGA, however, did not reflect the accuracy of analyses of transition, but their performativity. The HGA was an economic device (Callon et al. 2007: 2, Çalişkan and Callon 2009, Collier 2011) that was expected to enact the idea of transition in a practical form. The point was made explicit by BTC:

> the HGA followed a recognized approach that has been used for large-scale natural resources projects in developing countries and *economies in transition*. Project participants benefit from greater certainty to support their investment, and the host countries benefit from receiving investments that may not otherwise be made, revenues that would not otherwise be generated and undertakings from project participants, which in the absence of the HGAs would not otherwise exist under national legislation. (BTC/RR 2003: 38, emphasis added)

Indeed, the idea that the HGA pointed towards the future was correct. Georgian labour law had already been liberalised during the 1990s, and following the Rose Revolution it became liberalised still further during the period of the Saakashvili government. The HGA did look forward to the future of Georgian labour law in practice.[27]

Yet, in principle, the idea of an exceptional legal regime could have been considered problematic, for along with others clauses in the HGA, the clause potentially contradicted the general principles of the OECD guidelines on multinational enterprises, which stated that companies should 'refrain from seeking or accepting exemptions not contemplated in the statutory or regulatory framework related to environmental, health, safety, labour, taxation, or other issues' (OECD 2000: 19). In Turkey, as we have seen, Amnesty International, along with other international NGOs (Amnesty International 2003, The Corner House et al. 2011), had addressed the question of the exception, arguing that the original wording of the HGA with Turkey led to the effective exclusion of the route of the pipeline from the ongoing development of human rights legislation in Turkey (see Chapter 2). However, the UK national contact point for the implementation of the OECD guidelines later agreed with the company that BTC had not 'sought or accepted exemptions not contemplated in the statutory or regulatory framework' (UKNCP 2011: 1).

As a number of commentators have noted, neo-liberalism does not take a single standardised form (Barry et al. 1996, Larner 2000, Ong 2007). And if the clause on labour legislation in the HGAs had a neo-liberal logic, displacing state legislation with an agreement to respect customary industry practice, it was nonetheless apparently tempered by ethical undertakings. For despite the absence of public information about working conditions, there was one aspect of employment that was raised in public documents by both BTC and the IFIs. This was the issue of local employment, for, as one of its social objectives, pipeline subcontractors were expected to employ workers from affected communities along the route of the pipeline (BTC/ESM 2007: 22). For some pipeline workers, this requirement was double-edged. On the one hand, some unskilled and semi-skilled workers living in the vicinity of the pipeline were employed in construction projects in a region of high unemployment and widespread poverty (BTC/ESIA 2002d: 11-6).[28] On the other hand, the workers were therefore employed on short-term contracts as construction work passed through their area, enabling them to be dismissed easily.[29] Moreover, some workers had to pay bribes to intermediaries in order to gain employment in the first place, even when they were to be employed on short-term contracts. While, as we have seen, the archive was discreet about the potential for corruption in state institutions, the problem of bribery in the labour market had been acknowledged, as consultation had 'repeatedly suggested that there was potential for corruption and/or bias in the recruitment process' (ibid.: 11-7) and led the company to make this public in Georgia. The point was made by senior BTC management in a broadcast on national media: 'Do not pay any groups or individuals offering to mediate between you and BP and Spie Capag Petrofac [the contractors]' (BTC 2004). While the HGA bypassed the provisions of the Georgian labour code, the oil company had, in effect, introduced its own ethical labour code by insisting on the need to employ workers from particular regions. At the same time, in stating their opposition to corruption to the Georgian public, it sought to maintain the fragile border between the transparent economy of the pipeline and the exterior of this economy. The limits of what was made public had to be adjusted in response to a changing political situation.

Multiple Histories

The BTC pipeline was designed not only to be a technical means for the transportation of oil, but a distinct space of economic intervention, the constitution of which depended on the use of a set of economic as well as technical devices. The Host Government Agreements provided the legal basis on which this space could be constructed and purified as distinct from other elements of the economy of Azerbaijan, Georgia and Turkey; the ESIA and RAP mapped the territory of economic intervention in advance of its realisation in practice; and the expectation was that the publication of the price levels used for compensation would enable the process to be transparent and fair. This was a device that was intended to render both things (land), and behaviours and processes (labour), in an economic form (Callon et al. 2007: 3, Mitchell 2008), thereby apparently forging a distinction between the transparency of the pipeline's corridors and the wider economy and society within which they were located.

In practice, however, it proved to be a more complex task to build such an autonomous and transparent space of economy than it was to imagine one. To understand why specific disputes occurred one would need to analyse the way in which the economic interventions of the company co-existed and interfered with a series of other interventions. History, and its continuing material presence, mattered. Further analysis of the specific political situations that developed along the route of the pipeline would have to address the customary use of land, the accuracy of land registry records, the history of land privatisation (cf. Verdery 2003, Verdery and Humphrey 2004, Yalçin-Heckmann 2010), the extent of fraud (Schueth 2012, Green Alternative 2005), the migration of Greeks, Svans, Ajara and refugees from Abkhazia (Trier and Turashvili 2007), the importance of beekeeping and fruit farming in rural areas, and the collapse of industrial and agricultural production in Georgia in the aftermath of the break-up of the Soviet Union (Lerman 2006: 116). It would also need to address the expectations fostered by the Shevardnadze government (Chapter 5), the location of military bases (Chapter 2), the operation of the community investment programme, the weakness of the trade unions, and the strength of civil society organisations in Georgia (Hamilton 2004). At the same time, as we have seen, the movement of trucks, pipes, soil and water enlarged the pipeline's corridors and blurred their borders (see also Chapter 6). These elements all contributed to the formation of the new regime that was being established along the length of the pipeline. The pipeline corridor was not a stable enclosed system, a transparent enclave, distinct from the post-Soviet economic life beyond its limits (Chapter 5). On the contrary. To understand the disputes that emerged along the route of the pipeline we have to attend to the shifting, contested and uncertain boundaries of, as well as the complex interplay between, what was rendered transparent and what was not.

Chapter Nine
Conclusions

Geopolitical Fieldwork

In August 2008 war broke out between Georgia and Russia over the region of South Ossetia, which had been autonomous from the rest of Georgia since the civil war of the early 1990s. While there has been a great deal of debate about the causes of the conflict and the objectives of its protagonists, some commentators have suggested that the Russian intervention was not just directed at the presence of Georgian forces in South Ossetia or at the pro-Western regime of Mikheil Saakashvili, but also at the BTC pipeline, which had been completed only a few years earlier.[1] The politics of the pipeline had been dominated in the mid-2000s by concerns about the social and environmental impact of construction work; but when, as part of the South Ossetian conflict, bombing occurred in the area of the pipeline, it appeared to demonstrate that the pipeline continued to be of wider geopolitical importance. Bertrand Russell's supposition about the link between Russian invasions of Georgia and the oil economy was thus reproduced.

There is, however, some doubt about whether Russian forces had sought to bomb the BTC pipeline or not, or indeed whether the conflict had anything to do with the pipeline at all (House of Commons 2009: 64).[2] One expert informant in Tbilisi reckoned that the bombs might have been aimed at the pipeline, but missed due to the inaccuracy of Russian bombing; another thought the bombs were probably directed at Georgian military

Material Politics: Disputes Along the Pipeline, First Edition. Andrew Barry.
© 2013 John Wiley & Sons, Ltd. Published 2013 by John Wiley & Sons, Ltd.

Figure 9.1   Energy Infrastructure, Akhali Samgori, September 2010. Photo taken by author

vehicles as they dispersed away from a nearby military base.[3] Moreover, there were good reasons for thinking that the Russian air force did not intend to damage the pipeline for, if this had been the intention, it would have made sense to target one of the pumping stations that were clearly visible from the air, rather than the pipeline, which was not. BP was reported to have stated that there was 'no evidence' that the pipeline had been bombed.[4]

While, as we have seen, the locations of the environmental impacts of pipeline construction are recorded copiously in company documents, it is difficult to determine the location of any bomb craters in the vicinity of the pipeline. When an assistant and I set out to visit the bomb craters with the help of local residents, lacking any precise information, it took most of one afternoon to determine their location – or at least that of one set of craters – after navigating the potholed roads of the partially derelict and heavily polluted industrial zone of Rustavi, a city that had been a major centre for steel production in the Soviet Union. Helped by a villager and his son, we eventually found four craters lined up in a neat row near a small river lying approximately one kilometre from the village of Akhali Samgori, north-east of Rustavi and possibly 400–500 metres from the BTC pipeline itself. Our guides told us that an Azeri woman who had been working in the fields on the other side of the river had died of a heart attack when the bombs exploded, but there were no other casualties. At the time I did not know that two researchers had already carried out extensive fieldwork in the area, reaching the conclusion that Russian planes had not targeted the BTC pipeline but that they had dropped at least 45 bombs near the Baku-Supsa pipeline (Marriott and Minio-Paluello 2012: 151, Chapter 4).[5] In Akhali Samgori we found, however, that villagers were convinced that the bombs had been targeted at the BTC pipeline. Yet they were much more concerned about the lack of energy in the village itself. While the BTC pipeline was now operating as an underground artery of oil, rusting empty gas pipes ran along the sides of the streets several metres above the ground. The village was dominated by the presence of both Soviet and post-socialist energy infra-structures, but derived little benefit from the existence of either.[6] If, during the 1990s, political analysts thought that the relations between the BTC pipeline and geopolitical interests were clear enough, the case of Akhali Samgori reminds us that it may be as difficult to decipher the material traces of geopolitics as the material consequences of environmental impacts.

The Limits of Transparency

At Akhali Samgori the pipeline is now fully functioning and buried. Its route is marked out by a series of yellow signs that warns landowners and farmers of the presence of the pipeline and the risks that it might pose; while for those not living immediately in its vicinity, its existence continues to be made visible through the ongoing publication of information about its impact and operation (Chapter 6). As we have seen, this has resulted in the publication of a huge public archive of documents, which largely forms the basis for this book.

I suggested earlier that practices of transparency and corporate social responsibility appear to offer capital a progressive and enlightened way of deal-ing with potentially disruptive actions. BTC's transparency, as we have seen, was expected both to meet demands for greater accountability and to foster the 'free exchange of ideas' (Chapter 3). The publication of project documents was intended to facilitate informed and reasonable debate about issues that concerned affected populations, ranging from pollution, employment oppor-tunities and compensation to the protection of the archaeological heritage, thereby enabling the oil company to address and manage the problems that mattered. Problems could be anticipated, discussed and addressed before they escalated to the levels of passionate and violent conflict that have often been associated with the operation of the oil economy both elsewhere and in the past. One of the objectives guiding the principle of transparency was, above all, to ensure peace (cf. Shapin and Schaffer 1985, Toscano 2007).

However, to understand the politics of the pipeline, I have suggested in this book, it is necessary to attend both to what is made transparent in the growing archive of public documents – its presences, so to speak – and to the limits of transparency – what might appear as significant absences. From the accounts in preceding chapters, I want to bring out four such presences and absences in the public documentation. The first arises from the manner in which the state figures in the archive. On the one hand, there are frequent references to the importance of the state. We learn, for example, of the requirements set out in the environmental permit granted by the Georgian government, which highlighted the importance of security against sabotage (Chapter 2), and of the consequences of weaknesses in the earlier land registration process. But, on the other hand, we learn little about the politics of the land privatisation that happened only a few years earlier, nor about villagers' everyday experience of the state in the settlements along the pipeline route (Chapter 8). And if the practice of transparency is expected to act as an antidote to the opacity of the state, the published documents remain largely silent about the nature, the prevalence and the causes of the myriad social and economic problems that transparency was expected to address (Chapter 3). The presence of the 'state' in all of its manifestations is marked primarily by its positioning in the margins of published documents.

A second absence from the archive is more systematic and revolves around employment practices and the politics of labour. The organisation of relations between BTC and its various contractors was considered largely a matter for 'internal monitoring'; at the same time, the information published in the archive had little to say about the pay and working conditions of workers or the occurrence of strikes and stoppages. If factory inspectors once produced rich accounts of the conditions in nineteenth-century British factories, providing empirical descriptions that served as the basis for an analysis of exploitation, no equivalent public record exists of labour conditions along the route of the pipeline (Chapter 8). The transnational politics of BTC revolved incessantly around questions of the environmental and social responsibilities of the company, but these were defined and bounded in such a way as not to include the politics of labour (Chapter 4). Of course, this does mean that the politics of labour in connection with the pipeline were not registered in Georgia, for they were a minor facet of national politics in the wake of the Rose Revolution; it is just that they did not figure in the transnational political debate. In turn, this indicates again how the transparency of BTC catalysed and intensified the interest of transnational observers in very particular events, objects and issues, while effectively marginalising others.

A third marker of the limits of transparency concerns an imbalance in the provision of certain kinds of scientific information and research.[7] The archive publicises abundant quantities of information about the movement and distribution of materials; at the same time, detailed accounts of primary research are less available and appear only in highly mediated and reduced forms. The documents that are made public include such matters as accounts of the location of pipe storage and construction yards, the distribution of endangered species and archaeological sites, and the precise location of river crossings. But the archive goes further than this. It includes annually updated indicators of problems such as noise, air pollution and ground and surface water contamination, as well as traffic accidents, injuries and fatalities to pipeline workers (BTC/ESAR 2008: 54). It even includes astute criticisms of the progress of the project and accounts of the obstacles that it encountered (Chapters 6 and 8), thus manifesting a considerable degree of institutionalised reflexivity on the part of the company. Writers on governmentality have long been attentive to the methods used to regulate the conduct of persons; but the archive presents us with a wealth of evidence on how a corporation sought in addition to govern the existence, activity and movement of materials, as well as the effects that these materials could generate. If, as I have argued, materials lie at the heart of the politics of the pipeline, then this reflects the content of the archive.

Yet although the archive turns our attention repeatedly towards the significance of materials, there are evident limits to the accounts that it provides and the transparency it effects. For in general, the published documents do not contain the original research reports written by those charged with investigating problems such as landslides, vibration or pollution, and tell us little about their research practices. Researchers' reports are not, in general, made public (Chapter 7); and if the conclusions of such reports are made public, they are summarised in or translated into accessible, 'non-technical' language for those who are immediately affected (Chapter 6). In this way, the form and degree of scrutiny of research reports that could in principle be exercised by external experts and other observers is managed and contained. Moreover, a key feature of the disputes that arose was the paucity of independent sources of research and expertise existing external to the apparatus of scientific and technical monitoring established by BTC and its lenders (Chapters 2 and 6). In this light, a number of the disputes that emerged along the pipeline route are particularly intriguing in so far as they direct us to consider the conduct of research itself, as well as the importance of documents that were not published in the archive. Thus, while the activities of the state and the politics of labour are marginalised in published documents, the archive also remains reticent about the research and ensuing reports and publications that were centrally implicated in mediating many of the controversies that arose (cf. Jasanoff 2006b, McGoey 2007). This is not to say that original research reports were not circulating unofficially, but they did not have the status of being officially part of the archive.

But a fourth element is also apparent in the management of 'presences' in the archive. This centres on the constitution of the pipeline route as a series of corridors each of which governed specific aspects of its construction, operation and impact – from issues of safety and compensation to environmental impacts. For this pronounced spatialisation of information production created borders between the informational spaces, borders that were themselves ambiguous and contestable, and which therefore became key sites for fomenting dissent (Chapters 5, 6 and 8).

A core argument running through this book is that, if transparency might be expected to foster informed and rational debate while limiting the scope and intensity of controversy, this does not occur as anticipated. For as the case of BTC demonstrates, the production of information – in the form of the evolving archive – had the effect of multiplying the surfaces on which disagreement can incubate and flourish. This took two forms. The publication of information, in the guise of archival 'presences', generated a series of questions and problems about which it was possible to disagree. But the boundaries between what was present and what was significantly absent from the archive also fostered dissent. Consider the ways in which the limited figuring of the state in the archive became a focus for distrust. In distinctive ways, critical observers in Azerbaijan, Turkey and Georgia noted weaknesses in published accounts of the relation between the pipeline project and the state as shown by the coverage of such issues as land ownership, corruption and security. How was it possible to trust what had been made public, when so much was evidently not (Chapters 2, 3, 6 and 8)? Consider, too, the presences and absences manifest in the published scientific information on environmental impacts. Instead of limiting the possibility for dispute, the publication of this information generated a host of new, potentially disputatious questions and problems. Disagreement grew over the very existence of the impacts, over their spatial limits, over how they might best be managed and mitigated, over the quality of the scientific information about impacts, as well as over environmental issues about which little scientific research had been published in the archive (Chapters 2 and 6). These dynamics generated a type of transnational politics in which the authority of experts became an intensely political matter. The reams of documentation provided, as I have noted, a series of additional and extended surfaces with which dissent could engage, and on the basis of which it could virally multiply. If one of the objectives of the archive is to establish a newly invigorated liberal political order based on the principle of transparency, then the archive registers the recurrent problems of how to address transparency's limits (cf. Mitchell 1999, Hibou 2011).

Material Politics

Transparency is expected to provide a way of reducing both the level and the intensity of the conflicts that have all too often marked the history of the oil industry, redirecting the politics of oil on to a more rational terrain. But oil companies have to manage the unruly behaviour of materials as well as the disruptive actions of persons. As we have seen, the BTC archive provides us with an audit of the company's efforts to govern and monitor the activity of materials. Its contents register the company's response both to the multiple demands of international guidelines and principles and to the threat that public criticism poses to its reputation. But the quantity of scientific information contained in the archive is equally an index of the range of the technical challenges posed by the construction of a pipeline. Within the confines of laboratory experiment it may be possible, to a greater or lesser extent, to regulate the activity of materials or to design their properties and internal structure, isolating them in relatively pure forms (Chapter 7). But in contrast, a pipeline cannot be purified in this way: it forms part of a lively and dynamic assemblage of materials and persons, the activity of which it is arduous and costly to govern. It exists in an evolving and contested environment, its integrity potentially threatened by landslides, neglect, sabotage, corrosion or the failure of particular material components. Indeed, the case studies collected together in this book register in diverse ways how difficult it may be to monitor the potentially disruptive behaviour of materials in the field. Such occurrences as the movement of land or the fragility of pipes demonstrate the limits of the capacity of the company to govern this activity, creating the possibility for controversy to erupt.

My claim is therefore not that materials have political agency in themselves, nor that materials have political significance in general. This book does not offer a general account of the role material agency plays in political life (cf. Bennett 2010). Instead, my approach to the study of material politics is guided by a commitment to a certain form of empiricism, one that requires us to attend at once to the specificity of materials, to the contingencies of physical geography, the tendencies of history and the force of political action. The political significance of materials is not a given; rather, it is a relational, a practical and a contingent achievement.

To illuminate and extend this stance, over the course of this book I have examined a series of apparently distinctive disputes that sprang up at specific points along the route of the BTC pipeline concerning matters ranging from landslide prediction to the practice of public consultation. I have argued, however, that such disputes cannot be treated simply as a series of discrete cases and should be understood as elements of political situations. I have done so for three reasons. First, as I suggested in Chapter 4, the significance and extent of such specific disputes is in principle perspectival and underdetermined. For the BTC company and the IFIs, such controversies were both resolvable and ultimately resolved. They were amenable to technical solutions and, in this way, did not immediately raise any wider questions, although they did ultimately generate 'lessons' (IFC 2006). In contrast, radical critics sought to establish connections between the disparate controversies that occurred along the pipeline and beyond, directing us towards their resonances with events elsewhere. Although some controversies revolved around apparently minor issues, when considered cumulatively and collectively they became for these critics signs of more systemic problems.

A second reason why we need a concept of the political situation is illustrated by the case studies presented in Chapters 5 to 8. In these chapters we saw that, on the one hand, multiple sites in which controversies have emerged can come to be linked together as part of the same causal nexus. The disputes surrounding the alleged failure of the company to consult and inform the residents of particular villages along the route is an example of this process (Chapter 5). On the other hand, we saw also how individual disputes may come to be elements of multiple knowledge controversies or political situations. Indeed, certain disputes may become particularly inflamed due to the contingent interference in them of a number of otherwise unrelated political situations, diffracted in a single site. The intensity of the controversy surrounding the village of Dgvari is a striking example of this phenomenon associated with political situations (Chapter 6). The controversy over Dgvari was a dispute that came to be figured as an element in a number of political situations: among them the limits of corporate social responsibility, the impact of landslides, the capacity of the Georgian state and the politics of the Borjomi route.

A third reason for using the term political situation concerns the importance of what C.S. Peirce termed the logic of abduction. As I have argued, abduction is one key mechanism at work in establishing a link between a specific dispute, or set of disputes, and a wider political situation. For Peirce, abduction was a form of reasoning or intuition that moved from the apprehension of a 'surprising fact' to a cause that might explain this fact. In Chapter 4 I proposed that the practice of abduction makes it possible both to conjure up and to transform the context within which a particular problem or series of problems can be explained. Collectively, the set of disputes along the pipeline generated for critics a surprising fact: that despite its espousal of the values of corporate social responsibility and transparency, BTC failed to enact these values in practice (Chapters 4 and 5). The very enactment of a political situation – conjured up through the act of abduction – served itself to contribute to and transform the political situation.

In light of these observations we can see why some of the disputes that I have discussed were never fully resolved. Technically and procedurally they were resolved, and the resolution to these disputes is recorded in the archive. In general, either corporate experts were able to arrive at a solution to a problem, or they argued that there was not really any impact or issue that needed addressing at all. In this way, corporate scientific expertise ultimately had anti-political consequences, closing down the apparent grounds for and the possibility of further disagreement about a specific issue (Barry 2002). However, as we have seen, environmental and social problems often have multiple causes and many elements, which cannot readily be addressed through technical solutions alone. What are taken to be impacts are abstractions from the events of which they are elements (Chapter 6). In contrast to the experts, critics called attention to the voices of the villagers, their anxieties about environmental risks and their anger about the lack of compensation they would receive (Chapter 8). And they took specific disputes to be indices of much wider problems, not just particular technical issues. In short, the corporation did not try to engage with the series of wider questions that the critics attempted to raise, while the critics generally did not concern themselves with the technical details of the corporation's solutions to the problem of impacts, or all the iterations of multiple layers of monitoring and reporting. The evidence provided by scientific research was challenged and countered by evidence generated through political fieldwork, and the authority of science was placed in opposition to the experience and observation of non-experts drawing on non-expert testimony. This impasse is a familiar one to those concerned with the politics of science. What is unusual in this case, however, is that transparency was proposed as a solution to this impasse, with the expectation that it would foster a more consensual and a more informed politics.

If we accept the democratic value of agonistic dissensus rather than either consensus or antagonism, then two conclusions follow. Certainly, we need to recognise the uncertainties, necessary abstractions, simplifications and absences as well as the professional ethics and judgements that are routine features of scientific research. Rather than allow for disputes about specific problems to be closed down on the basis of the authority of science, there is a need both to support and to value the possibility of disagreement about the results of research. However, just as we should acknowledge the inevitable uncertainties and limitations of scientific research, we should also attend to the uncertainties and simplifications of politics. In thinking about the politics of oil, there is a tendency to understand this politics too readily in geopolitical or economic terms, overdetermining the significance of specific controversies and events. In these circumstances one of the tasks of social research is to disturb any sense of the self-evidence of this political context.

Coda

It is a warm day in late September 2010, shortly before the start of the Tbilisi international contemporary art festival, Artisterium.[8] I am accompanying the artist Mamuka Japharidze along the route of a gas pipeline which threads its way down the steep hill from his house to the city of Tbilisi below. He tells me that this was the route along which the Red Army came into Tbilisi ninety years previously. I had already learned that during the construction phase of BTC, Mamuka had decided to paint the words 'This is not a pipe' on the Tbilisi gas pipeline. The painted words had since faded or been erased, but he had previously exhibited photographs of his temporary artwork in an exhibition in Baku.

Japharidze's homage to René Magritte poses – just as Magritte's painting does – the question of how one might understand the relation between words and objects. Do Magritte's words, as Foucault (1983) once asked, refer to an object (a pipe), the painting of an object (clearly not a pipe itself), the relation between different pipes (in Magritte's later painting of his original painting), or the words themselves and their interrelation ('this', after all, is not a 'pipe')? Japharidze's work raises some of these questions too, but differently. The Tbilisi gas pipeline, unlike the Magritte pipe, clearly is a pipe of a kind, although it is not the BTC pipeline, on which the artwork was intended, in part, to be a commentary. But Japharidze's writing also points to a more salient truth, namely, that the BTC pipeline itself is much more than a material object – a pipe – in the first place. While probably nothing had been written about Magritte's pipe before Magritte painted his famous surrealist axiom, the construction of the oil pipeline has always been bound up with accounts of what it is. For when the oil company developed the pipeline it created an object which became associated with an extraordinary apparatus of information production, leading to the generation of a public archive of commentary, monitoring, auditing and public criticism, as well as the circulation of documents between Tbilisi, Baku, London and beyond. 'This is not a pipe' because the pipe to which Japharidze's work obliquely but clearly refers was always much more than steel, although it is that too.

Notes

1 Introduction

1 The International Finance Corporation, a member of the World Bank Group, supports private sector investments in developing countries. The IFC loan amounted to $300 million out of what was then projected to be $3.7 billion construction costs

2 In this book, I use the term information very broadly. I take it to refer to any written account of a plan, procedure, legal agreement, social and environmental assessment or scientific data. In this context, information is expected to inform others, but its production also transforms the object of the information. As I have argued previously, information has three characteristics: 1) the transformation of the object about which information is produced; 2) 'the [anticipated] transformation in the conduct of those who are, or should be, informed'; and 3) the existence of a multitude of regulatory arrangements, technical standards and institutional resources that enable information to be produced (Barry 2001: 153–154).

3 Studies of knowledge controversies initially focused primarily on controversies that occurred between scientists (e.g. Collins 1981). More recent studies have been increasingly concerned with controversies that, although they may involve scientists, take place within a wider public realm. I term these latter controversies, public knowledge controversies.

4 One should not imagine that the local, the regional and the global amount to a hierarchical series of scales (Amin 2004, Allen 2004, Allen and Cochrane 2007, Powell 2007). Local processes are always likely to be irreducible to the wider social context within which they are so often framed (Tarde 1999 [1893]).

5 As Mariam Motamedi-Fraser argues, 'it is impossible to draw up a list of entities that enter an event in advance because identities and relations acquire definition through it' (Fraser 2010: 65).

Material Politics: Disputes Along the Pipeline, First Edition. Andrew Barry.
© 2013 John Wiley & Sons, Ltd. Published 2013 by John Wiley & Sons, Ltd.

6   An analytics of the situation, then, is concerned to highlight a nexus of different historical movements, material processes, interests, ideas and practices, brought together in novel and shifting conjunctures or configurations, and leading to unanticipated effects (cf. Whitehead 1920, Fraser 2006). Aspects of Marx's analysis of politics (Marx 1973 [1852]), Tarde's analysis of invention and imitation (Tarde 2001, Barry and Thrift 2007, Born 2010), Gramsci's critique of economistic readings of Marx (Gramsci 1994, Jessop 2008), and the work of more recent political theorists such as Hannah Arendt, Bernard Crick, and Chantal Mouffe, all provide accounts, in different ways, of the irreducible complexity of political situations (Crick 1962, Arendt 1963, Honig 1993, Mouffe 1993).

7   During the early stages of the project the archive had, for a short period, an identifiable physical presence. In the summer of 2002 it existed in a printed form in a small number of locations, including the offices of the European Bank for Reconstruction and Development in London and Baku. It was in this printed form that the film-maker and environmentalist, Martin Skalsky, located documents in Baku and Yevlakh during the period of the 120-day consultation process established by the EBRD and IFC (see Chapter 5). The archive also became accessible on a website hosted by BP and dedicated to the project along with other oil projects in the Caspian region. While this site has since disappeared, its contents can still be found on the website of BP in the Caspian where it is identified as 'Environmental and Social Documentation'.

8   'Our task is not to give voice to the silence that surround, as [statements], nor to rediscover all that, in them and beside them, had remained silent or reduced to silence. Nor is to study the obstacles that have prevented a particular discovery, held back a particular formulation, repressed a particular form of enunciation, a particular unconscious meaning or a particular rationality in the course of development; but to define a limited system of presences' (Foucault 1972: 119).

9   This antagonism was palpable in certain events that influenced my decision to undertake research on the BTC pipeline. I first became aware of BTC in the autumn of 2002. I had been thinking about how to develop a research project on the politics of branding and reputational management that might complement the work of my colleague Celia Lury on branding (Lury 2004, Power 2007a). A Greenpeace worker whom I had interviewed about their Stop E$$O campaign directed me to the launch of a book about BTC (Platform et al. 2003) and recommended the work of the environmental and social justice art organisation Platform, who were known for their research on the oil industry. At the launch event at the UK House of Lords, a succession of speakers documented the human rights abuses and environmental damage associated with the activities of BP in Columbia, Alaska and the North Sea. The book presented research carried out by a coalition of international NGOs along the pipeline route. This raised immediate questions. How could a multinational oil company and international NGOs, based in London, generate facts about the social and environmental consequences of a 1760km pipeline running across a region that, in the West, was poorly understood? What series of transformations had occurred between the work of the oil company consultants, IFI specialists and NGO researchers in the field, and the divergent accounts that they each came to produce eventually? At this time, I was particularly intrigued by the idea of a study of the practice of what I have come to call political fieldwork.

2   The Georgian Route: Between Political and Physical Geography

1   On Mackinder's own brief participation in British intervention in southern Russia in 1919 see Kearns 2009: 202–213.

2   Nonetheless, there had been some limited British interest in Georgian oil: 'two wells were ... begun ... near Chatma by a British company, but during the Tartar-Armenian riots they were destroyed and work remained in abeyance until 1916, when boring was taken up again and pushed to 660 feet, where the oil was met' (Ghambashidze 1919: 61).

3   In this respect, the position of Georgia might be compared with that of the Lebanon, which was a key transit route for oil from the Middle East. On the relation between oil transportation and US intervention in the Lebanon in the 1940s and 50s, see Gendzier 1997.

4   The Mensheviks had briefly led the government of an independent Georgia from1918–21 prior to its incorporation in the Soviet Union. Kautsky himself was impressed by what he termed the 'social experiment' of the Menshiviks, and visited Tbilisi from September 1920 until January 1921. Having remained in the capital he was modest about his understanding of the country as a whole: 'Thus I cannot pose as one who has investigated the country. Nevertheless, I have learned far more of it than an ordinary tourist; everybody most readily gave me information upon all things that I asked about; both the heads of the Government and officials as well as the representatives of the Opposition; proletarians as well as business people and intellectuals' (Kautsky 1921, preface to the English edition).

5   O'Tuathail and Dalby include studies of 'techniques of governmentality' in their account of the domain of critical geopolitics (1998). However, critical studies of geopolitics have, in practice, not been concerned primarily with the role of the natural sciences as techniques of governmentality.

6   This view is contested by Lincoln Mitchell, who persuasively argues that the Saakashvili government captured the US administration as much as vice versa (L. Mitchell 2009).

7   A consortium of ten international oil companies, including BP, Chevron, TPAO, and Statoil, formed to develop Caspian oil resources.

8   Walker suggests that 'the key difficulty broached by claims about globalization is that the modern political imagination has always expressed considerable ambivalence as to whether our political situation is grounded in the territorial (though politically constituted and not simply natural) spaces of particular states, or in some apparently more abstract realm, in some world in which we can be more at home with our humanity' (Walker 2010: 88).

9   Interviews, Washington, DC and the Department for International Development, London, March–October 2004.

10   'Akhalkalaki Residents Rally Against the Pullout from Russian Base', *civil.ge*, 13 March 2005. The base eventually closed in 2007; see N. Landrau, 'The Evacuation of the Russian Military Base Comes to a Close', *Caucaz Europenews*, 30 May 2007.

11   Appendix 3 of routing report.

12   Maia Chalaganidze, 'If Nothing Changes, We Will Begin a Serious Campaign Against BP', *24 Saati*, 28 November 2002.

13   The National Movement became increasingly prominent during the latter stages of the Schevardnadze government, particularly following the local elections of 2002, after which Saakashvili became chairman of Tbilisi City Council (Nodia and Scholtbach 2006: 19).

14   Kote Kemularia comment in *24 Saati*, 28 November 2002, see also Nikoloz Rurua, 'The Great Protagonists of Oil and Society', *24 Saati*, 28 November 2002. Rurua criticised the government for labelling anyone critical of the pipeline as an enemy of the state.

15   Giorgi Gakheladze, 'The Environment Ministry Will Not Sacrifice Borjomi', *24 Saati*, 30 November 2002.

16   'Fate of BTC is Still Undecided', http://www.civil.ge/eng/article.php?id=2804, 30 November 2002; 'Georgia Wants Alternative Routes of BTC', http://www.civil.ge/eng/article.php?id=2806, 30 November 2002; 'Political Agenda Overshadows Environmental Concerns', http://www.civil.ge/eng/article.php?id=2811&search=, 2 December 2002; C. Smith, 'Baku-Ceyhan: The Geopolitics of Oil', http://www.opendemocracy.net/democracy-caucasus/pipeline_2763.jsp, 16 August 2005 (accessed May 2013); Green Alternative 2005.

17   Comment in *24 Saati*, 28 November 2002.

18   'Go-ahead Given to BTC', www.civil.ge, 2 December 2002; 'Environmental Protests Linger After Pipeline's Approval', www.eurasianet.org 17 December 2002 (accessed May 2013).

19   'Georgia Won't Be Intimidated', *Eurasianews.net*, 5 August 2004, available at http://www.eurasianet.org/departments/recaps/articles/eav080604.shtml (accessed May 2013).

20   Interview, Tbilisi, October 2004.

21   The World Bank had been financing a programme to develop appropriate areas of expertise in Georgia since 2001. However, within the World Bank group there was a 'firewall' between this $9.6 million programme (World Bank 2001) and IFC finance for the BTC pipeline. In practice, the programme did not fund the development of expertise in Georgia but was used to pay Western consultancy firms to assist the Georgian government. However, funding for this programme could not be used in relations of security, which remained the responsibility of the State (interview, Washington, DC, November 2004).

22   Saeed Shah, 'Rumsfeld Intervention Rescues $3bn BP Pipeline', *Independent*, 5 August 2004.

23   Ibid.

24   'BTC Co Grants $46 Million to Georgia', www.civil.ge, 11 October 2004 (accessed May 2013); Green Alternative 2005.

25   I return to consider the critical importance of this archive to the emergence of disputes along the pipeline in Chapters 5 and 8.

26   http://www.equator-principles.com (accessed May 2013).

27   Tony Juniper, Friends of the Earth, quoted in O. Bowcott, 'Unstable Artery', *Guardian*, 23 July 2003.

28   http://www.indymedia.org.uk/en/2002/12/48040.html?c=on#c48177 (accessed May 2013).

29   Oliver Balch, 'Principles in the Pipeline', *Guardian*, 8 December 2003.

30   Fieldwork and interviews, London, November 2003–March 2004.
31   The following year, protestors handed out a mock company annual report with a picture of Tony Blair in military uniform on the front cover to shareholders arriving at the Annual General Meeting of BP. In this way, the protest reframed the political situation in macropolitical terms.
32   Interviews, EBRD, London 2003–04.
33   http://www.bakuceyhan.org.uk/missions.htm (accessed May 2013).

3   Transparency's Witness

1   http://www.indymedia.org.uk/en/2002/12/48040.html?c=on#c48177 (accessed May 2013).
2   One response to this limitation is to gather more information and associate EITI with a wider range of interventions. This is the approach, termed 'EITI++' proposed by the World Bank. '[This] will provide governments with a slate of options including technical assistance and capacity building for improving the management of resource-related wealth for the benefit of the poor. Through technical assistance, EITI++ aims to improve the quality of contracts for countries, monitoring operations and the collection of taxes and royalties. It will also improve economic decisions on resource extraction, managing price volatility, and investing revenues effectively for national development'. World Bank press release 2008/269/AFR, Washington, DC 12 April 2008.
3   http://eiti.org/TimorLeste (accessed May 2013).
4   That is, the fund set up to receive revenue payments. A substantial fraction of payments made by oil companies to the Azerbaijan government takes the form of tax, which is not recorded.
5   Interview, Baku, June 2004.
6   Interview, Baku, June 2004
7   Interview, Baku, June 2004.
8   For an account of the history of corruption and organised crime in Georgia during the post-Soviet period, see Kukhianidze (2009).
9   Interviews and field notes, Tbilisi, March 2004, Baku, June 2004.
10   The guidelines contain numerous references to the need for transparency as a means of preventing corruption and bribery, for example: http://www.oecd.org/daf/inv/mne/48004323.pdf (accessed May 2013).
11   The Equator Principles are described as, 'A financial industry benchmark for determining, assessing and managing social and environmental risk in project financing'. Those institutions that adopt the principles, 'recognise the importance of transparency with regard to the implementation of the Equator Principles (EPs)' http://www.equator-principles.com/ (accessed May 2013).
12   In principle, it would be possible to trace some of the negotiations and compromises that have led to this particular body of international agreements and laws – in other words, to demonstrate its relation to politics. Yet in practice firms, investors, international institutions and NGOs tend to treat this evolving regulatory constellation as the non-political foundation which governs, but does not determine, what should or should not be rendered transparent. Nonetheless, the provisional quality of such guidelines and laws means that the

distinction between law and politics may be hard to sustain. In this respect, this body of law and soft law appears quite different from the specific case of administrative law analysed by Bruno Latour (2009). On the relation between legal and political disagreement, see Waldron (1999b).

4 Ethical Performances

1 On environmental and social impact assessment, see Becker 1997, Barrow 2000. The construction of the Trans-Alaska oil pipeline was critical to the development of the idea and practice of environmental impact assessment. The practice of social impact assessment developed later, and came to be become increasingly important in the 2000s.

2 The Brent Spar was referred to as a buoy by Royal Dutch Shell, but as an installation, facility, platform, floating infrastructure or rig by other observers including the press. Part of the difficulty in classifying the Brent Spar was that it was both floating and yet also massive, immobile and part of a network of other massive material artefacts including rigs and tankers. As a floating object, the Brent Spar could be described as a buoy, but as a massive element of an oil production and transportation infrastructure, it might be described as a facility or installation. While one term (buoy) points to the individuality of the Brent Spar as a physical object, the notion of facility points to its existence as part of a much larger network of objects.

3 My reading of abduction is much indebted to Gell (1998).

4 The BTC and IFI documents refer to the village as Garabork.

5 Of course, my own interpretation and abductive inferences are also entering into the political situation via this book.

5 The Affected Public

1 BTC/ESIA 2003, Article 5.2. In addition the project was expected to conform to the terms of the Espoo Convention on Environmental Impact Assessment in a Transboundary Context and the Convention on the Protection and Use of Transboundary Watercourses and International Lakes.

2 Interviews and fieldwork, 2003–4. Whereas BP employed a chief economist, it largely outsourced social research to external firms and universities.

3 Of course, it would be a mistake to imagine that publics form in the way that they are expected or intended to, or that publics are preformed possibilities that are somehow brought into existence or assembled through use of a particular technique, whether it is a public consultation meeting or an opinion poll. As Vikki Bell argues, instead of thinking that subjects are performed, 'the alternative process ... is a process of actualization guided by difference and creation. Rather than the realm of the possible (and the real), one has virtuality (and actualization)' (Bell 2007: 107). In this Deleuzian formulation, publics can be understood as virtualities, actualised in more or less novel forms, whether through simulation or imitation, or through unanticipated and creative acts of re-invention. In this chapter I focus not on the unanticipated acts of the population, but on the claims that were made by others to speak on its behalf.

4   See Baku-Ceyhan Campaign (2003a&b) and IFC (2003a). Susan Barker notes that the Åarhus convention, 'seeks to promote sustainable development through granting procedural rights. Such rights include citizen access to information, the right to public participation and access to justice in environmental matters. It is premised on the belief that granting procedural rights will enable citizens to participate directly in environmental decision making, thereby enhancing the quality of environmental policy … [but it] has a rather limited notion of participation, stressing the primacy of representative institutions and giving a constrained role to public participation' (Barker 2006: 115). According to the principles of the 1998 Åarhus convention, public authorities should ensure that if there is any request for environmental information by the public, 'copies of the actual documentation containing or comprising such information' should be made available (UNECE 1998, Article 4, 1). Following the decision by IFIs to support the construction of the BTC oil pipeline, in principle, the process of public consultation took this further. Information concerning the pipeline was to be made available without a request being made, and concerning matters which went far beyond the realm of the 'environment', specified in the convention.

5   For example, 'a "one-off cash payment" made directly to the affected household was the overwhelming preference (96 percent) of survey respondents for compensation payment' (BTC/RAP 2002d: 4-32).

6   Interviews and fieldwork in Georgia March–October 2004.

7   In the terms of MacIntosh and Quattrone's elegant analysis of management accounting and control systems, the ESIA was expected to perform the role of a learning machine, informing the corporation how it should act (MacIntosh and Quattrone 2010: 331).

8   Space was not just produced, but marked out in advance in order to be reconfigured (Lefebvre 1984, Foucault 2002b).

9   The four inch difference was an adaptation to the environmental and political sensitivity of the Georgian section, enabling BTC to construct fewer pumping stations in Georgia than if a narrower diameter pipe had been used.

10   In Turkey the ESIA was termed an EIA according to Turkish law. In practice, the EIA did include social impact assessment in Turkey.

11   Interview, Ankara, September 2004.

12   Interview, London, November 2003.

13   For example, the *Guardian*, 3 September 2002, www.newint.org, www.earthfirst. org.uk, www.bankwatch.org.

14   Social Impact Assessment could be described not so much as a scientific discipline but as a community of practice, 'a system of relationships between people, activities and the world; developing with time, and in relation to other tangential and overlapping communities of practice', such as environmental impact assessment (Lave and Wenger, quoted in Amin and Roberts 2008: 11). The expression 'social impact assessment' is said to have come about initially as a result of the controversy surrounding the construction of the trans-Alaskan oil pipeline in the 1970s, drawing on the provisions of the 1969 US National Environmental Policy Act. The act called 'for federal agencies to make integrated use of the natural and social sciences to prepare an environmental impact statement' (Barrow 2000: 9-10). The idea of social (as well as environmental) impact

assessment gained support from the 1980s following the publication of the Brundtland report on sustainable development (1987), the Rio Earth Summit (1992), the wave of 'anti-globalisation' protests of the late 1990s, along with the growing influence of international NGOs on the policies of international financial institutions, including the World Bank (Burdge 1995, Burdge and Vanclay 1995, Becker 1997, Barrow 2000, Noble 2010).

15 Marriott's account of the case of Haçibayram proceeds as follows:

> The study was published in June 2002. Its careful detail reassures readers in the cities that time and attention have been given in the villages and fields. Its little 'T' marked on the map by Haçibayram indicates that here the consultation with villagers was carried out by telephone. Yet all the time the study was under way the village was in ruins. There were no telephones. There was no-one to answer them.
>
> As Mehmet climbs along the roof among the hives, that 'T' lies buried in computers and CDs in the offices of ERM and BP. It is as though a huge funnel had sucked up the fields and the hay harvest, the ruins and the evictions, and concentrated them into one tiny byte of information. The noise of the bees, the breeze of the late afternoon, is translated into some new language, so that the eyes of those few who read it, glowing on a computer screen, can say 'yes!' (Marriott 2003)

As we have seen earlier, a distinctive feature of Marriott's work is his use of a variety of literary and artistic forms in order to express the experience, the wider significance and the complexities of what I have termed political fieldwork. In the case of Haçibayram he used the form of a fable, echoing an argument made by John Law, who explicitly draws the connection between sociology and literary and artistic practice: 'In Euro-America the inscriptions that condense ontic/epistemic imaginaries belong to the novel or to poetry or to art and not to serious research method. As do those that condense *non-coherences* (James Joyce?), *overpowering fluxes* (Edvard Munch?), *indefiniteness* (Mark Rothko? Franz Schubert?), *multiplicities* (Georges Braque?) or *fractionalities* (Steve Reich?)' (Law 2004: 148).

16 The routing of the pipeline near the town of Borjomi, and near to the National Park, had been one of the key issues in the broader debate surrounding the construction of the pipeline (see Chapter 2). The region of Borjomi was considered environmentally sensitive because of the existence of the National Park, the prevalence of landslides in the area and because the town of Borjomi was famous as a source of mineral water. The WWF had been active in the area since the early 1990s and adopted a hard line against the Borjomi route. For the WWF's position, see, for example, 'WWF Alarmed at World Bank Impotence on BTC Pipeline Decision', November 2003, http://www.wwf.org.uk/wwf_articles.cfm?unewsid=676 (accessed May 2013). WWF was not merely concerned with environmental issues but also with the problem of reputational risk: 'Any oil pollution even if it does not get into the aquifers will destroy the reputation of the Borjomi Water Industry affecting 1000s of jobs in the region. The Borjomi tourism industry will be severely damaged, as will the German government's plans to help develop tourism, improvements to water supplies, sewage, waste etc (EUR 10 million facilitated by WWF)' (WWF 2003: 31).

17 Interviews and fieldwork in Baku, Tbilisi, Borjomi and Kars, 2004.

6   Visible Impacts

1   Indeed, those short sections of the pipeline route that would remain visible throughout its lifetime – pumping stations – were initially assigned individual Community Liaison Officers. By contrast, those Community Liaison Officers who were assigned to cover 'invisible' sections of the route were given long stretches. Presumably it was anticipated that the location of pumping stations – the visible markers of the pipeline – would have greater impact and therefore were more likely to be controversial. In practice, the pumping stations were relatively uncontroversial; rather, disputes proliferated along those sections of the pipeline that would ultimately be invisible. As a result more Community Liaison Officers had to be recruited to cover these sections.

2   Along with producing a report on the practice of public information disclosure, Martin Skalsky, together with his colleague Martin Maraček, made a short documentary film in Azerbaijan called *The Source*. The film, which was not widely distributed, has the sense of a black comedy. In it, Skalsky and Maraček interview a senior manager of the State Oil Company of Azerbaijan. The manager says very little, but talks briefly and ironically of the earlier history of Soviet films about the Caspian oil industry, with images of heroic workers. In an ironic act of post-socialist solidarity with the Czech film-makers, he allows his 'comrades' from the former Soviet Union to film in the oil fields of Azerbaijan. Scenes from this interview along with shots taken from archive Soviet films are juxtaposed with the polluted environment of old onshore oil fields. When the film-makers speak to a senior BP manager, however, he tells them 'there is nothing for you to see'. The new BTC pipeline had already been buried underground and rendered invisible. The implication was that in a world in which everything has been made public or disclosed, direct observation of materials and places is both impossible, but also unnecessary.

3   Interviews Baku, March 2004 and Washington, DC.

4   Whereas BP managers oversaw the development of the pipeline in Georgia and Azerbaijan, Botaş took a lead role in Turkey. On the broader question of Turkey's continual 'failure' and efforts to measure up to European standards, see Ahiska 2003. In her ethnographic study of the post-socialist Polish meat-packaging industry, Elizabeth Dunn traces the difficulty for Polish meat producers in meeting European health and environmental standards. Dunn shows that the prospective entry of Poland into the EU led to a division between those firms that could become something like European firms and those which developed alternative markets (to Russia and the Ukraine) where European standards did not apply. The difference between earlier socialist factories and the factories that met European standards is that the latter became marked by the disciplinary power of audit rather than older forms of direct surveillance. 'Managers' disciplinary gaze', she notes, 'thus becomes more powerful, not less, as it is mediated by paper logs' (Dunn 2005: 185).

5   Although I refer to such impacts as 'measured impacts' I recognise that such impacts are the subject of both quantitative and qualitative analysis (Callon and Law 2005).

6   A number of studies of environmental politics have documented the ways in which efforts to frame issues as environmental problems are often in tension with other understandings (e.g. Kropp 2005) and 'key alternative local framings' (Fairhead and Leach 2003: 106).

7 According to Dgvari residents, September 2010.

8 Interviews, Tadzrisi and Tbilisi, September 2004.

9 While not specifically referring to the issue of vibration, the BTC external specialist monitoring panel (SRAP) recognised that construction traffic was often the cause of minor disputes: 'In Georgia, disputes with local communities have arisen where the construction contractor has temporary access roads without adequately confirming land ownership or consulting with adjacent villages. Where these incursions lie beyond designated rights of way ... it can lead to conflict with adjacent communities and leaves the project vulnerable to criticism that it has not complied with OD 4.30 [the World Bank operational directive on voluntary resettlement]' (BTC/SRAP 2004a: A-11).

10 The following is based on field visits to Sagrasheni in June and October 2004 and September 2010. In the early stages of the project in Georgia, the company had not anticipated the range of difficulties and conflicts, over land ownership and construction work, which would develop between the company, contractors and villagers. This led to the recruitment of further Georgian-speaking community liaison officers in 2003–4.

11 'A 79-year-old single woman from Sagrasheni ...witnessed heavy trucks passing close to her house for the last four years. According to the project documentation the trucks should use a different route, and yet every day during the construction period BTC trucks not only brought pipes for the construction site, but they also used the same road to take the concrete blocks from the Tetritskaro cement factory to other construction sites. Living on a pension and sustaining herself with her fruit and vegetable garden, she "understands" the importance of the BTC pipeline for the country; the only question she needs an answer to is why her house has started to crack since 2003?'(CEE Bankwatch 2006: 69).

12 According to Mol, hospital doctors may be concerned when there is a marked difference between the disease that is presented in an interview with the patient and that which is made present through a physical examination (Mol 2002: 51).

13 While there is a substantial literature on the relationship between vibration and the body, much of it focuses on the somatic experience of sound (e.g. Nancy 2007: 76–77).

14 Interview, Tbilisi, June 2004; Kirtadze 2005; fieldwork September 2004, September 2010.

15 *24 Saati*, 24 January 2004.

16 'Villagers Protest Against BTC Construction', www.civil.ge, 12 January 2004 (accessed May 2013); 'Police Break Up Anti-BTC Rally', www.civil.ge, 21 August 2004 (accessed May 2013).

17 The distinction between those events that were mediated by international institutions and those associated with direct action, however, is not clear cut. The Netherlands Commission on Environmental Impact Assessment noted that some villagers, frustrated by the speed of formal processes, also took part in blockages. Residents of the city of Rustavi both engaged in public protests and filed a complaint to the IFC demanding compensation. Workers from the city also engaged in unofficial strike action.

7 Material Politics

1   In this respect, this chapter follows others which argue for the need to over-turn the conventional hierarchy of the disciplines which places 'fundamental' sciences (physics, molecular biology and the neurosciences) at the top of the hierarchy, and less fundamental disciplines, including chemistry, agronomy, metallurgy, physical geography and social anthropology further down (Schaffer 2003, Stengers and Bensaude-Vincent 2003, Barry 2005).

2   In the chapter I leave aside the question of how the relation between metallurgy and the broader field of materials science is conceived by actors. Metallurgy, along with materials science more broadly, is in any case an interdisciplinary field which incorporates elements of chemistry, physics, crystallography and, indeed, management theory. On the broader question of the interdisciplinarity of disciplines see Barry, Born and Weszkalnys 2008.

3   Whitehead uses the example of the mountain to explain endurance as a process of transformation: 'The mountain endures. But when after ages it is worn away, it has gone' (Whitehead 1985: 107).

4   'In industry, it is rarer to see a "materials department", rather technical departments will now tend to be identified by the product – or in the aerospace sector as the "system" or "platform". An aeroengine is a system in this sense, and the technical team will involve materials scientists alongside aerodynamicists, structural engineers, electrical engineers, designers, etc. [In] university research, we are moving slowly to this systems approach, or "interdisciplinary" research as it is more normally called in the academic sector. Many of the modern challenges in materials are not solely about "new" materials, but rather materials' integration into systems with specified overall function.' P. Grant, personal communication, 2007.

5   ECGD provided up to $150 million cover for the project (House of Commons 2004–5: 9).

6   The Corner House describes itself as a group which aims to support democratic and community movements for environmental and social justice through research and advocacy. Its approach is based on evidence: 'we try to take a "bottom-up" approach, filled with examples, to issues of global significance which are often handled in a more abstract way', www.thecornerhouse.org.uk (accessed May 2013).

7   Fieldnotes, April 2004.

8   The Georgian Green Movement had been founded as early as 1988 (Wheatley 2005: 48). One of its first leaders, Zurab Zhvania, was Prime Minister (2004–5) in the Saakashvili government. In comparison to Georgia, political interest in environmental issues is undeveloped in neighbouring countries including Azerbaijan and Turkey.

9   On the role of mediators see Osborne 2004.

10 As Bruno Latour notes, the word 'fact' means something quite different in science and the law: 'rather than confuse the two, we should sharpen the contrast: when it is said that the facts are there, or that they are stubborn, that phrase does not have the same meaning in science as it does in law, where, however stubborn

the facts are, they will never have any real hold on the case as such, whose solidity depends on the rules of law that are applicable to the case' (Latour 2004: 89). While the operation of a select committee has some similarities to a court of law it is a distinct form of political assembly, the characteristics of which have yet to be investigated.

11 The work of the engineer is an indicator of the complex geography of knowledge production in the oil industry which relies on the production of a whole series of different forms of knowledge, which may be more or less attuned to the existence of local specificities (Bridge and Wood 2005: 206).

8 Economy and the Archive

1 In other words, the international financial institutions (EBRD and IFC), a number of national export credit departments and investment banks, and 'any other export credit agencies and commercial lenders and other providers of debt financing or political risk insurance for the BTC project' (BTC/IEC 2004: 5).

2 i.e. content supplied by local firms. The need to monitor local content reflects the criticism that the contracts relating to major infrastructure investments such as BTC primarily benefit international companies.

3 In Georgia, the work of the NGOs formed part of the Pipeline Monitoring and Dialogue Initiative (PMDI). The NGOs involved did not include Green Alternative, which had taken an explicitly oppositional stance to the company and the IFIs. There was, nonetheless, a question of how independent and how critical those involved in PMDI could be of the project (interview, Tbilisi, September 2010).

4 The internal levels were called 'BP and BTC shareholder monitoring', 'BTC project monitoring and operations management', 'BTC assurance team', and 'construction contractors' (BTC/ESR 2004b: 13).

5 In Georgia this objective had four specific elements: 'to improve income-earning and economic opportunities; to support the development and improvement of the agricultural sector; to improve living conditions through rehabilitation of social infrastructure; to improve the capacity of communities to self-organise, manage and self-initiate community driven development' (BTC/PCIP 2003: 186).

6 'Mercy Corps' overall strategy in Georgia is to strengthen Georgian Civil Society by promoting sustainable and equitable socio-economic development involving civic groups, government and the private sector while emphasizing the leading role of local communities. During CIP-E, Mercy Corps plans to work with local communities as they address their development priorities in such a way as to leave them more confident and competent at the end of CIP-E', http://www.mercycorps.org/countries/georgia/10204 (accessed May 2013).

7 From my fieldnotes, Tbilisi and Krtsanisi, June 2004.

8 Interview, Tbilisi, September 2010.

9 Interviews, Tbilisi, 2004.

10   One report from an international NGO fact-finding mission argued that there
were shortcomings in the CIP. However, the report does not necessarily point
to the failure to perform specific actions, but to the potential for confusion
between ethical investment, formal compensation for losses in agricultural
production, and the informal payments in kind performed by subcontractors in
order to smooth the progress of construction work. The mission quotes a
villager from the western Georgian village of Sakuneti in this way: 'Because the
pipeline is crossing our land there was a promise to make good things happen
in the village. But nothing good has happened, apart from the renovation of the
club in the village. The village wanted the club to be renovated but the company only renovated the inside; and they promised to renovate the roads, but
nothing has happened' (CEE Bankwatch et al. 2004: 30). It is not clear from
this account whether the CIP had actually failed to deliver or there was a
confusion between the actions of the company [in promising to renovate roads]
and community investment [the renovation of a club or ritual house]; it is also
unclear whether 'the company' referred to BP or the BTC subcontractors or
the NGO commissioned to perform the CIP.

11   While there appear to have been few problems with CIP projects in Georgia, a
few projects in Turkey engendered complaints. A report from one international
NGO fact-finding mission to Turkey contended that although an extensive programme of artificial insemination for cattle was heavily promoted by an NGO
entrusted to manage projects funded by CIP, the programme had largely been
a failure. Indeed, according to the FFM 'having lost 70–80% of a year's calves,
the economic costs are major: all the villages estimated the figure to be in the
thousands of dollars' (Centre for Civic Initiatives et al. 2005b: 25). A subsequent report by the BTC SRAP panel, carrying out fieldwork in the same area,
came to the opposite conclusion, recording that 'all villagers with cattle were
able to participate and the villages consulted spoke of the very positive benefits
and success rate of these programmes' (BTC/SRAP 2006b: D-16).

12   Interview, Tbilisi, April 2004, see also BTC/SRAP 2005a: C-22–23.

13   A shift in the location of the route near to Tadzrisi occurred in order to reduce
the impact of landslides, interview, Borjomi, June 2004.

14   Interview, Tbilisi, June 2004.

15   Interviews and fieldwork, Georgia, 2004.

16   For example, 'BP/BTC then provided a satellite photograph of the plot following
pipeline construction. Although the features are poorly defined in this latter photograph, it is still possible to identify images of 3 or 4 trees adjacent to the pipeline
corridor (located to the right of the corridor in the satellite photo). Further,
whereas the English translation of the land inventory for the above plot indicated
"illegible" regarding the number and description of trees on the plot, the original
Georgian version of the inventory indicated 3 trees were included in the inventory'
(EBRD/IRM 2009: 16).

17   A term used by one informant working for the BTC company.

18   From my field diary, June 2004.

19   This 300m rule was nonetheless contested by some Georgian beekeepers, who
argued that the distance should not be determined by the impact of construction work but by the flight distance of bees which could, in mountainous areas,
be as far as 13.5 km.

20 According to the World Bank, vulnerability 'denotes a condition characterized by higher risk and reduced ability to cope with shock or negative impacts. It may be based on socio-economic condition, gender, age, disability, ethnicity, or other criteria that influence people's ability to access resources and development opportunities. Vulnerability is always contextual, and must be assessed in the context of a specific situation and time.' http://go.worldbank.org/HSXB13LCA0 (accessed May 2013).

21 From my notes of meetings with Georgian Trade Union leaders in Tbilisi and pipeline workers in Rustavi, March 2004.

22 Information provided by the construction contractor, Tbilisi, 2004.

23 *Alia*, 18–19 December 2003, cited in Green Alternative et al. 2004.

24 'The majority of local mayors and people we speak [with] are concerned with the employment issue and situation [and] that pipeline does not support any other business development in the region. In [the] Tetritskaro region, people express disappointment with the fact that [the] company is bringing from Tbilisi everything including food, beverage[s], and water. There is no region where new business activities or development [have] been fixed because of pipeline construction. [L]ots of the people complain about [the] untransparent process of selection of the workers. It should be mentioned that the expectations of the people [concerning the number of] ... jobs are very high' (Green Alternative 2003: 3)

25 Interviews, Tbilisi and Baku, April 2004.

26 A similar statement was made in the Resettlement Action Plan (BTC/RAP 2002g: 3–4). Nonetheless one Georgian lawyer concerned with the development of BTC thought that the clause meant that the Georgian labour code still governed pipeline construction (GYLA 2003). This was also the view of other Georgian informants.

27 Interview, Tbilisi, September 2010.

28 The practice of employing local workers could lead to tensions between ethnic groups. In the Tsalka region in central Georgia, which had a mixed Armenian, Greek and Azeri population with a Georgian minority, the prospect of employment on the BTC project encouraged internal migration from Adjara and Svaneti, exacerbating existing tensions between local residents and the newcomers who occupied abandoned Greek villages in the area. In these circumstances, the construction of the pipeline became drawn into a conflict between local Armenians and Greeks, on the one hand, and Georgian migrants, on the other: '... although the BP-led consortium was committed to employing local people in the construction of the pipeline, with priority given to those living within 2 km of the pipeline or within 5 km of Above Ground Installations, in Tsalka district it was unclear what "local" meant, given the ongoing influx of migrants from other parts of the country. For whatever reason, it turned out that the local staff employed on the pipeline were mainly ethnically Georgian migrants from Adjara and Svaneti. In contrast, relatively few members of the Armenian community found work on the project and young Armenian men continued to travel to Russia for seasonal labour' (Wheatley 2006: 24).

29 According to the trade unions the employers tended to hire workers on contracts of no more than three months while stating that the first three months were 'probationary' (notes from meeting with Georgian trade unions, March 2004).

9   Conclusions

1   *Daily Telegraph*, 11 August 2008, The Corner House et al. 2008, Carroll 2012: 293, On the Russian-Georgian war of 2008 see especially Cornell and Starr 2009. On the recent history of Russo-Georgian relations see especially Gordadze 2009. The pipeline had been attacked in eastern Turkey just prior to the start of the conflict in South Ossetia, *Oil and Gas Journal*, 6 August 2008. However, the causes of this attack are disputed (House of Commons 2009, Marriott and Minio-Paluello 2012).

2   *Daily Telegraph*, 10 August 2008, *Daily Telegraph*, 13 August 2008, *Guardian*, 11 August 2008, 'Fears Over Stability Over Georgian Pipeline' *Spiegel Online*, 13 August 2008.

3   Fieldwork, Tbilisi, September 2010, *International Herald Tribune*, 14 August 2008.

4   *International Herald Tribune*, 14 August 2008.

5   During the war, a Reuters' report noted that three bombs craters were seen near the nearby Baku-Supsa pipeline http://uk.reuters.com/article/2008/08/29/uk-georgia-ossetia-pipelines-idUKLT53399920080829 (accessed May 2013).

6   Fieldwork, September 2010. On post-socialist energy infrastructures see, in particular, Humphrey (2003) and Collier (2011).

7   I am grateful to Irem Kok, whose research stresses the importance of this point.

8   http://artisterium.org/Artisterium_V3/ (accessed May 2013).

References

Abbott, K. and Snidal, D. (2000) 'Hard and Soft Law in International Governance', *International Organization*, 54, 421–456.

Adams, T. (1999) 'Oil and Geopolitical Strategy in the Caucasus', *Asian Affairs*, 30, 1, 11–20.

Agrawal, A. (2005) *Environmentality: Technologies of Government and the Making of Subjects*, Durham, NC: Duke University Press.

Ahiska, M. (2003) 'Occidentalism: The Historical Fantasy of the Modern', *South Atlantic Quarterly*, 102, 2/3, 351–379.

Ahiska, M. (2010) *Occidentalism in Turkey: Questions of Modernity and National Identity in Turkey*, London: I.B. Tauris.

Ahlbrandt, T.S. (2006) 'Global Petroleum Reserves, Resources and Forecasts', in R. Mabro (ed.) *Oil in the Twenty-First Century*, Oxford: Oxford University Press, 128–173.

Akiner, S. (ed.) (2004) *The Caspian: Politics, Energy, Security*, London: Routledge Curzon.

Allen, J (2004) 'The Whereabouts of Power: Politics, Government and Space', *Geografiska Annaler*, 86B, 19–32.

Allen, J. and Cochrane, A. (2007) 'Beyond the Territorial Fix: Regional Assemblages, Politics and Power', *Regional Studies*, 41, 9, 1161–75.

Althusser, L. (1984) 'Ideology and Ideological State Apparatuses', in *Essays on Ideology*, London: Verso.

Amin, A. (2004) 'Regions Unbound: Towards a New Politics of Place', *Geografiska Annaler*, 86B, 33–44.

Amin, A. and Roberts, J. (eds) (2008) *Community, Economic Creativity and Organization*, Oxford: Oxford University Press.

Amin, A. and Thrift, N. (2005) 'What's Left: Just the Future', *Antipode: A Radical Journal of Geography*, 37, 2, 220–238.

Amirahmadi, H. (ed.) (2000a) *The Caspian Region at a Crossroad: Challenges at a New Frontier of Energy and Development*, Basingstoke: Palgrave.

Material Politics: Disputes Along the Pipeline, First Edition. Andrew Barry.
© 2013 John Wiley & Sons, Ltd. Published 2013 by John Wiley & Sons, Ltd.

Amirahmadi, H. (2000b) 'Pipeline Politics in the Caspian Region', in H. Amirahmadi, *The Caspian Region at a Crossroad: Challenges at a New Frontier of Energy and Development*, Basingstoke: Palgrave.

Anderson, B. (2006) 'Becoming and Being Hopeful: Towards a Theory of Affect', *Environment and Planning D*, 24, 5, 733–752.

Anderson, B. (2007) 'Hope for Nanotechnology: Anticipatory Knowledge and Governance of Affect', *Area*, 19, 156–165.

Anderson, K. and Braun, B. (eds) (2010) *Environment: Critical Essays in Human Geography*, Aldershot: Ashgate.

Andolina, R., Laurie, N. and Radcliffe, S. (2009) *Indigenous Development in the Andes; Culture, Power and Transnationalism*, Durham, NC: Duke University Press.

Appel, H. (2011) 'Offshore Work: Oil and the Making of Modularity in Equitorial Guinea', unpublished PhD thesis, Stanford University.

Arendt, H. (1963) *Eichmann in Jerusalem: a Report on the Banality of Evil*, New York: Viking.

Asdal, K. (2008) 'On Politics and the Little Tools of Democracy. A Down to Earth Approach', *Distinktion. Scandinavian Journal of Social Theory*, 16, 5–26.

Ashenden, S. (2004) *Governing Child Sexual Abuse: Negotiating the Boundaries of Public and Private, Law and Science*, London: Routledge.

Auty, R. (1993) *Sustaining Development in Mineral Economies: The Resource Curse Thesis*, London: Routledge.

Auty, R. (2005) 'Natural Resources and Civil Strife: a Two-Stage Process' in P. Le Billon (ed) *The Geopolitics of Resource Wars*, London: Routledge, 29–49.

Bakker, K. (2003) *An Uncooperative Commodity: Privatising Water in England and Wales*, Oxford: Oxford University Press.

Bannon, I. and Collier, P. (eds) (2003) *Natural Resources and Violent Conflict*, Washington, DC: The World Bank.

Barker, S. (2006) *Sustainable Development*, London: Routledge.

Barnett, C. (2003) *Culture and Democracy: Media, Space and Representation*, Edinburgh: Edinburgh University Press.

Barrow, C.J. (2000) *Social Impact Assessment: An Introduction*, London: Hodder Headline.

Barry, A. (1999) 'Demonstrations: Sites and Sights of Direct Action', *Economy and Society*, 28, 1, 75–94.

Barry, A. (2001) *Political Machines: Governing a Technological Society*, London: Athlone.

Barry, A. (2002) 'The Anti-political Economy', *Economy and Society*, 31, 2, 268–284.

Barry, A. (2004) 'Ethical Capitalism', in W. Larner and W. Walters (eds) *Global Governmentality*, London: Routledge.

Barry, A. (2005) 'Pharmaceutical Matters: The Invention of Informed Materials', *Theory, Culture and Society*, 22, 1, 51–69.

Barry, A. (2006) 'Technological Zones', *European Journal of Social Theory*, 9, 2, 239–253.

Barry, A. (2010) 'Tarde's Method: Between Statistics and Experimentation', in M. Candea (ed.) *The Social after Gabriel Tarde: Debates and Assessments*, London: Routledge, 177–190.

Barry, A. (2012) 'Political Situations: Knowledge Controversies in Transnational Governance', *Critical Policy Studies*, 6, 3, 324–336.

Barry, A. and Born, G. (2013) *Interdisciplinarity: Reconfigurations of the Social and Natural Sciences*, London: Routledge.

Barry, A. and Thrift, N. (2007) 'Gabriel Tarde: Imitation, Invention and Economy', *Economy and Society*, 36, 4, 509–525.

Barry, A., Born, G. and Weszkalnys, G. (2008) 'Logics of Interdisciplinarity', *Economy and Society*, 37, 1, 20–49.

Barry, A., Osborne, T. and Rose, N. (eds) (1996) *Foucault and Political Reason: Liberalism, Neo-Liberalism and Rationalities of Government*, London: UCL Press.

Barthe, Y. (2006) *Le Pouvoir d'Indécision: le mise en politique des déchets nucléaires*, Paris: Economica.

Barthes, R. (1973) *Mythologies*, London: Palladin.

Bayulgen, O. (2009) 'Caspian Energy Wealth: Social Impacts and Implications for Regional Stability', in A. Wooden and C. Stefes (eds) *The Politics of Transition in Central Asia and the Caucasus: Enduring Legacies and Emerging Challenges*, London: Routledge, 163–188.

Beck, U. (1992) *Risk Society: Towards a New Modernity*, London: Sage.

Beck, U. (1999) *World Risk Society*, Cambridge: Polity.

Becker, H. (1997) *Social Impact Assessment*, London: UCL Press.

Behrends, A., Reyna, S. and Schlee, G. (eds) (2011) *Crude Domination: An Anthropology of Oil*, Oxford: Berghahn.

Bell, V. (2007) *Culture, Performance: The Challenge of Ethics, Politics and Feminist Theory*, Oxford: Berg.

Bennett, J. (2005) 'In Parliament With Things', in L. Tønder and L. Thomassen (eds) *Radical Democracy: Politics Between Abundance and Lack*, Manchester: Manchester University Press, 133–148.

Bennett, J. (2010) *Vibrant Matter: A Political Ecology of Things*, Durham, NC: Duke University Press.

Bennett, T. and Joyce, P. (eds) (2010) *Material Powers: Cultural Studies, History and the Material Turn*, London: Routledge.

Bennie, L. (1998) 'Brent Spar, Atlantic Oil and Greenpeace', *Parliamentary Affairs*, 51, 3, 397–410.

Bennington, G. (1994) *Legislations: The Politics of Deconstruction*, London: Verso.

Bensaude-Vincent, B. and Stengers, I. (1996) *The History of Chemistry*, Cambridge, MA: Harvard University Press.

Bentham, J. (1839) 'On Political Tactics', in J. Browning (ed.) *The Collected Works of Jeremy Bentham part VIII*, Edinburgh: William Tait.

Bergin, T. (2011) *Spills and Spin: The Inside Story of BP*, London: Random House.

Berlant, L. (2007) 'On the Case', *Critical Inquiry*, 33, 4, 663–672.

Bernal J.D. (1969) *Science in History Volume 3: The Natural Sciences in our Time*, Harmondsworth: Penguin.

Best, J. (2005) *The Limits of Transparency: Ambiguity and the History of International Finance*, Ithaca, NY: Cornell University Press.

Best, J. (2007) 'Why the Economy is Often the Exception to Politics as Usual', *Theory, Culture & Society*, 24, 4, 83–105.

Billmeier, A., Dunn, J. and van Selm, B. (2004) 'In the Pipeline: Georgia's Oil and Gas Transit Revenues', IMF working paper WP/04/209, http://www.imf.org/external/pubs/ft/wp/2004/wp04209.pdf (accessed May 2013).

Bishop, C. (2012) *Artificial Hells: Participatory Art and the Politics of Spectatorship*, London: Verso.

Blanchard, O. (1997) *The Economics of Post-Communist Transition*, Oxford: Oxford University Press.

Blok, A. (2007) 'Experts on Public Trial: Democratizing Expertise through a Danish Consensus Conference', *Public Understanding of Science*, 16, 163–182.

Blok, A. (2010) 'Topologies of Climate Change: Actor-network theory, relational-scalar analytics, and carbon-market overflows', *Environment and Planning D: Society and Space*, 28, 5, 896–912.

Borch, C. (2012) *The Politics of Crowds: An Alternative History of Sociology*, Cambridge: Cambridge University Press.

Born, G. (2003) 'Strategy, Positioning and Projection in Digital Television: Channel Four and the Commercialisation of Public Service Broadcasting in the UK', *Media, Culture and Society*, 25, 6, 773–799.

Born, G. (2005a) *Uncertain Vision: Birt, Dyke and the Reinvention of the BBC*, London: Secker and Warburg.

Born, G. (2005b) 'On Musical Mediation: Ontology, Technology and Creativity', *Twentieth-century Music*, 2/1, 7–36.

Born, G. (2007) 'Future-making: Corporate Performativity and the Temporal Politics of Markets', in D. Held and H. Moore (eds), *Cultural Politics in a Global Age: Uncertainty, Solidarity and Innovation*. London: Oneworld, 288–296.

Born, G. (2010) 'On Tardean Relations: Temporality and Ethnography', in M. Candea (ed.) *The Social After Gabriel Tarde: Debates and Assessments*, London: Routledge, 230–247.

Born, G. (2011) 'Music: Ontology, Agency, Creativity', in L. Chua and M. Elliott (eds), *Material Agencies: Meaning and Mattering After Alfred Gell*. Oxford: Berg.

Bouzarovski, S. and Konieczny, M. (2010) 'Landscapes of Paradox: Public Discourses and Policies in Poland's Relationship with the Nord Stream Pipeline', *Geopolitics*, 15, 1, 1–21.

Bowden, T.P. and Tabor, D. (2001) *The Friction and Lubrication of Solids*, Oxford: Oxford University Press.

Bower, T. (2010) *The Squeeze: Oil, Money and Greed in the 21st Century*, London: HarperCollins.

Bowker, G. (1994) *Science on the Run: Information Management and Industrial Geophysics at Schlumberger, 1920–1940*, Cambridge, MA: MIT Press.

Bowker, G. and Star, S.L. (1999) *Sorting Things Out: Classification and its Consequences*, Cambridge, MA: MIT Press.

Boyer, P. (1994) *The Naturalness of Religious Ideas: A Cognitive Theory of Religion*, Berkeley, CA: California University Press.

Bradshaw, M. and Swain, A. (2004) 'Foreign Investment and Regional Development', in M. Bradshaw and A. Stenning (eds) *East Central Europe and the Former Soviet Union: The Post-Socialist States*, Harlow: Pearson.

Braun, B. (2008a) 'Theorising the Nature-Society Divide', in K. Cox, M. Low and J. Robinson (eds) *The Sage Handbook of Political Geography*, London: Sage, 189–204.

Braun, B. (2008b) 'Environmental Issues: Inventive Life', *Progress in Human Geography*, 32, 5, 667–679.

Braun, B. and Disch, L. (2002) 'Radical Democracy's "Modern Constitution"', *Environment and Planning D: Society and Space*, 20, 5, 505–511.

Braun, B. and Whatmore, S. (eds) (2010) *Political Matter: Technoscience, Democracy and Public Life*, Minneapolis, MN: Minnesota University Press.

Bridge, G. (2008) 'Global Production Networks and the Extractive Sector: Governing Resource-based Development', *Journal of Economic Geography*, 8, 3, 389–419.

Bridge, G. (2009) 'The Hole World: Scales and Spaces of Extraction', *New Geographies*, 2, 43–48.

Bridge, G. and Le Billon, P. (2012) *Oil*, Cambridge: Polity.

Bridge, G. and Perreault, T. (2009) 'Environmental Governance', in N. Castree, D. Demeritt, D. Liverman and B. Rhodes, *A Companion to Environmental Geography*, Oxford: Basil Blackwell.

Bridge, G. and Wood, A. (2005) 'Geographies of Knowledge, Practices of Globalization: Learning from the Oil Exploration and Production Industry', *Area*, 37, 2, 199–208.

Bromley, S. (1991) *American Hegemony and World Oil: The Industry, the State System and the World Economy*, University Park: Pennsylvania State University Press.

Browne, J. (2010) *Beyond Business*, London: Weidenfeld and Nicholson.

Buchan, J. (1993 [1916]) *Greenmantle*, Oxford: Oxford University Press.

Bulkeley, H. (2005) 'Reconfiguring Environmental Governance: Towards a Politics of Scales and Networks', *Political Geography*, 24, 8, 875–890.

Bulkeley, H. and Watson, M. (2007) 'Modes of Governing Municipal Waste', *Environment and Planning A*, 39, 2733–2753.

Burchell, G., Gordon, C. and Miller, P. (eds) (1991) *The Foucault Effect*, Chicago, IL: Chicago University Press.

Burdge, R.J. (1995) *A Community Guide to Social Impact Assessment*, Madison, WI: Social Ecology Press.

Burdge, R.J. and Vanclay, F. (1995) *Environmental and Social Impact Assessment*, Chichester: John Wiley & Sons, Ltd.

Burton, A. (ed.) (2005) *Archive Stories: Facts, Fictions, and the Writing of History*, Durham, NC: Duke University Press.

Burtynsky, E. (2009) *Oil*, London: Steidl.

Butler, J. (1993) *Bodies that Matter: On the Discursive Limits of 'Sex'*, London: Routledge.

Çalişkan, K. and Callon, M. (2009) 'Economisation Part 1: Shifting Attention from the Economy to Processes of Economisation', *Economy and Society*, 38, 3, 369–398.

Çalişkan, K. and Callon, M. (2010) 'Economisation Part 2: A Research Programme for the Study of Markets', *Economy and Society*, 39, 1, 1–32.

Callon, M. (ed.) (1998a) *The Laws of the Market*, Oxford: Basil Blackwell.

Callon, M. (1998b) 'An Essay on Framing and Overflowing: Economic Externalities Revisited by Sociology', in M. Callon (ed.) *The Laws of the Markets*, Oxford: Basil Blackwell, 244–269.

Callon, M. and Latour, B. (1981) 'Unscrewing the Big Leviathan: How Actors Macrostructure Reality and how Scientists Help Them Do So', in K. Knorr-Cetina and A. Cicourel (eds) *Advances in Social Theory and Methodology*, London: Routledge and Kegan Paul, 277–303.

Callon M. and Law, J. (2005) 'On Qualculation, Agency, and Otherness', *Environment and Planning D: Society and Space*, 23, 5, 717–733.

Callon, M., Lascoumes, P. and Barthe, Y. (2001) *Agir dans un monde incertain: Essai sur la démocratie technique*, Paris: Seuil.

Callon, M., Millo, Y. and Muniesa, F. (eds) (2007) *Market Devices*, Oxford: Basil Blackwell.

Carroll, T. (2012) 'The Cutting Edge of Accumulation: Neoliberal Risk Mitigation, the Baku-Tbilisi-Ceyhan Pipeline and its Impact', *Antipode*, 44, 2, 281–302.

Carter, S. and McCormack, D. (2006) 'Film, Geopolitics and the Affective Logics of Intervention', *Political Geography*, 25, 228–245.

Chkheidze, I., Lominadze, G., French, D., Shilston, T., Whitbread, M., Lee, E., Morgan, D., Clarke, J. (2005) 'Geotechnical Engineering for Landslide Stability Management Along the Trans-Caucasus Oil and Gas Pipelines at Kodiana Pass – Sakire in Georgia', in M. Sweeney (ed.) *Terrain and Geohazard Challenges Facing Onshore oil and Gas Pipelines*, London: Thomas Telford, 428–445.

Cheterian, V. (2010) 'Azerbaijan', in D.O. Beachain (ed.) *The Colour Revolutions in the Former Soviet Republics: Successes and Failures*, London: Routledge, 101–117.

Clark, G. and Hebb, T. (2005) 'Why Should They Care? The Role of Institutional Investors in the Market for Corporate Global Responsibility', *Environment and Planning A*, 37, 2015–2031.

Clark, N. (2011) *Inhuman Nature: Sociable Life on a Dynamic Planet*, London: Sage.

Clarke, D. (2007) *The Battle for Barrels: Peak Oil Myths and World Oil Futures*, London: Profile.

Coles, K. (2004) 'Election Day: The Construction of Democracy Through Technique', *Cultural Anthropology*, 19, 4, 551–580.

Collier, P. and Hoeffler, A. (2005) 'Resource Rents, Governance, and Conflict', *Journal of Conflict Resolution* 49, 4, 625–633.

Collier, S. (2011) *Post-Soviet Social: Neo-Liberalism, Social Modernity, Biopolitics*, Princeton, NJ: Princeton University Press.

Collins, H.M. (ed.) (1981) 'Knowledge and Controversy', *Social Studies of Science*, 11, 1, 3–158.

Collins, H.M. (1985) *Changing Order: Replication and Induction in Scientific Practice*, London: Sage.

Collins, H.M. and Pinch, T. (1993) *The Golem: What You Should Know About Science*, Cambridge: Cambridge University Press.

Companjen, F. (2010) 'Georgia', in D. Ó. Beacháin and A. Polese (eds) (2010) *The Colour Revolutions in the Former Soviet Republics*, London: Routledge, 13–29.

Congress of the Georgian Soviets (1975) [1922] 'Manifesto of the Congress of Georgian Soviets to the Workers of the World', in L. Trotsky, *Social Democracy and the Wars of Intervention in Russia 1918–21*, London: New Park Publications, 111–115.

Cooke, B. and Kothari, U. (eds) (2001) *Participation: the New Tyranny?* London: Zed.

Corbridge, S., Williams, G., Srivastava, M., Véron, R. (2005) *Seeing the State: Governance and Governmentality in India*, Cambridge: Cambridge University Press.

Cornell, S. (2011) *Azerbaijan since Independence*, Armonk, NY: M.E. Sharpe.

Cornell, S. and Starr, F. (eds) (2009) *The Guns of August 2008: Russia's War with Georgia*, Armonk, NY: M.E. Sharpe.

Coronil, F. (1997) *The Magical State: Nature, Money, and Modernity in Venezuela*, Chicago, IL: Chicago University Press.

Corris, M. (1994) 'From Black Holes to Boardrooms: John Latham, Barbara Steveni, and the Order of Undivided Wholeness', *Art and Text*, 49, 66–72.

Crick, B. (1962) *In Defence of Politics*, London: Weidenfeld and Nicholson.

Croissant, M. (1999) 'Georgia: Bridge or Barrier for Caspian Oil?', in M. Croissant and B. Aras (eds) *Oil and Geopolitics in the Caspian Sea Region*, London: Praeger.

Cruikshank, B. (1999) *The Will to Empower: Democratic Citizens and Other Subjects*, Ithaca, NY: Cornell University Press.

Dányi, E (2011) 'Parliament Politics: A Material Semiotic Analysis of Liberal Democracy', unpublished PhD thesis, University of Lancaster.

Davies, G. (2006) 'Mapping Deliberation: Calculation, Articulation and Intervention in the Politics of Organ Transplantation, *Economy and Society*, 35, 2, 232–258.

Deffeyes, K. (2001) *Beyond Oil: The View from Hubbert's Peak*, New York: Hill and Wang.

Deleuze, G. (1979) 'Metal, Metallurgy, Music, Husserl, Simondon', Vincennes, 27 February, www.webdeleuze.com (accessed April 2013).

Deleuze, G. (1988) *Foucault*, London: Athlone.

Deleuze, G. (1993) *The Fold: Leibniz and the Baroque*, London: Athlone.

Deleuze, G. and Guattari, F. (1987) *Thousand Plateaus: Capitalism and Schizophrenia*, London: Athlone.

Demeritt, D. (2006) 'Science Studies, Climate Change and the Prospects for Constructivist Critique', *Economy and Society*, 35, 3, 453–479.

Derluguian, G. (2005) *Bourdieu's Secret Admirer in the Caucasus*, Chicago, IL: Chicago University Press.

Derrida, J. (1980) 'The Law of Genre', *Glyph*, 7, 202–32.

De Waal, T. (2004) *Black Garden: Armenia and Azerbaijan through Peace and War*, New York: New York University Press.

Dewey, J. (1927) *The Public and its Problems*, New York: Henry Holt.

Dewsbury, J.D. (2007) 'Unthinking Subjects: Alain Badiou and the Event of Thought in Thinking Politics', *Transactions of the Institute of British Geographers NS* 32, 443–459.

Dickson, L. and McCulloch, A. (1996) 'Shell, the Brent Spar and Greenpeace: a Doomed Tryst', *Environmental Politics*, 5, 1, 122–129.

Djelic, M.-L. and Quack, S. (eds) (2010) *Transnational Communities: Shaping Economic Governance*, Cambridge: Cambridge University Press.

Djelic, M.-L. and Sahlin-Anderson, K. (eds) (2006) *Transnational Governance: Institutional Dynamics of Regulation*, Cambridge: Cambridge University Press.

Dodds, K. (2003) 'Licensed to Stereotype: Popular Geopolitics, James Bond and the Spectre of Balkanism', *Geopolitics*, 8, 2, 125–156.

Dodds, K. (2007) *Geopolitics: A Very Short Introduction*, Oxford: Oxford University Press.

Dodds, K. and Sidaway, J. (2004) 'Halford Mackinder and the "Geographical Pivot of History": A Centennial Retrospective', *The Geographical Journal*, 170, 4, 292–297.

Dolan, C. and Rajak, D. (2011) 'Ethnographies of Corporate Ethicizing', *Focaal – Journal of Global and Historical Anthropology*, 60, 3–8.

Donald, J. (1992) *Sentimental Education*, London: Verso.

Dragadze, T. (2000) *Azerbaijan*, London: Melisende.

Driver, F. (2001) *Geography Militant: Cultures of Exploration and Empire*, Oxford: Blackwell.

Dryzek, J. (2000) *Deliberative Democracy and Beyond: Liberals, Critics, Contestations*, New York: Oxford University Press.

du Gay, P. (2000) *In Praise of Bureaucracy: Weber, Organisation, Ethics*, London: Sage.

Dunn, E. (2004) *Privatising Poland: Baby Food, Big Business, and the Remaking of Labor*, Ithaca, NY: Cornell University Press.

Dunn, E. (2005) 'Standards and Person-Making in East Central Europe', in A. Ong and S. Collier (eds) *Global Assemblages: Technology, Politics and Ethics as Anthropological Problems*, Oxford: Basil Blackwell, 173–193.

Durkheim, E. (1957) *Professional Ethics and Civic Morals*, London: Routledge.

Ebel, R. and Menon, R. (eds) (2000) *Energy and Conflict in Central Asia and the Caucasus*, New York: Rowman and Littlefield.

Eco, U. (1976) *A Theory of Semantics*, Bloomington, IN: Indiana University Press.

Eden, S. (1999) '"We Have the Facts" – How Business Claims Legitimacy in the Environmental Debate', *Environment and Planning A*, 31, 1295–1309.

Eden, S., Donaldson, A. and Walker, G. (2006) 'Green Groups and Grey Areas: Scientific Boundary-work, Nongovernmental Organisations, and Environmental Knowledge', *Environment and Planning A*, 38, 1061–1076.

Ehteshami, A. (2004) 'Geopolitics of Hydrocarbons in Central and Western Asia', in S. Akiner (ed.) *Caspian: Politics, Energy, Security*, London: Routledge Curzon, 55–67.

Elychar, J. (2005) *Markets of Dispossession: NGOs, Economic Development, and the State in Cairo*, Durham, NC: Duke University Press.

Engdahl, W. (2004) *A Century of War: Anglo-American Oil Politics and the New World Order*, London: Pluto.

Epstein, S. (1996) *Impure Science: AIDS, Activism and the Politics of Knowledge*, Berkeley, CA: California University Press.

Esposito, E. (2011) *The Future of Futures: The Time of Money in Financing and Society*, Cheltenham: Edward Elgar.

Fairhead, J. and Leach, M. (2003) *Science, Society and Power: Environmental Knowledge and Policy in West Africa and the Caribbean*, Cambridge: Cambridge University Press.

Featherstone, D. (2008) *Resistance, Space and Political Identities: The Making of Counter-Global Networks*, Oxford: Wiley-Blackwell.

Ferguson, J. (1999) *Expectations of Modernity: Myths and Meanings of Urban Life on the Zambian Copper Belt*, Berkeley, CA: University of California Press.

Ferguson, J. (2005) 'Seeing Like an Oil Company: Space, Security and Global Capital in Neoliberal Africa', *American Anthropologist*, 107, 3, 377–382.

Fischer, F. (2009) *Democracy and Expertise: Reorienting Policy Inquiry*, Oxford: Oxford University Press.

Fisher, E. (2008) 'The "Perfect Storm" of REACH: Charting Regulatory Controversy in the Age of Information, Sustainable Development and Globalization', *Journal of Risk Research*, 11, 4, 541–563.

Fisher, E. (2010) 'Transparency and Administrative Law: A Critical Evaluation', *Current Legal Problems*, 63, 272–314.

Fombrun, C. and Rindova, V. (2000) 'The Road to Transparency: Reputation Management at Royal Dutch Shell 1977–96' in M. Schultz, M. Hatch and M.H. Larsen (eds) *The Expressive Organisation: Linking Identity, Reputation and the Corporate Bond*, Oxford: Oxford University Press.

Forsythe, R. (1996) *The Politics of Oil in the Caucasus and Central Asia, Adelphi Papers*, London: International Institute for Strategic Studies.

Foster, H. (1995) 'The Artist as Ethnographer?' in G. Marcus and F. Myers (eds) *The Traffic in Culture*, Berkeley, CA: University of California Press.

Foucault, M. (1972) *The Archaeology of Knowledge*, London: Tavistock.

Foucault, M. (1979) *The History of Sexuality*, Harmondsworth: Penguin.

Foucault, M. (1983) *This is Not a Pipe*, Berkeley, CA: University of California Press.

Foucault, M. (2001) 'Les "reportages" d'idées', in *Dits et Écrits II, 1976–88*, Paris: Gallimard.

Foucault, M. (2002a) 'Questions of Method' in *Power: Essential Works: 1954–84*, London: Penguin, 223–238.

Foucault, M. (2002b) 'Space, Knowledge, Power' in *Power: Essential Works: 1954–84*, London: Penguin, 349–364.

Foucault, M. (2007) *Security, Territory, Population: Lectures at the Collège de France 1977–78*, Basingstoke: Palgrave Macmillan.

Franke, A. (ed) (2005) *Becoming Europe and Beyond*, Berlin: Kunst-Werke.

Fraser, M. (2001) 'The Nature of Prozac', *History of the Human Sciences*, 14, 3, 56–84.

Fraser, M. (2006) 'Event', *Theory, Culture and Society*, 23, 2–3, 129–132.

Fraser, M. (2010) 'Fact, Ethics and Event', in C. B. Jensen and K. Rodje (eds) *Deleuzian Intersections: Science, Technology, Anthropology*, Oxford: Berghahn, 57–82.

Frow, J. (2006) *Genre: The New Critical Idiom*, London: Routledge.

Fung, A., Graham, M. and Weil, D. (2007) *Full Disclosure: The Perils and Promise of Transparency*, Cambridge: Cambridge University Press.

Garsten, C. and de Montoya, G. (eds) (2008) *Transparency in a New Global Order: Unveiling Organizational Visions*, Cheltenham: Edward Elgar.

Gell, A. (1998) *Art and Agency*, Oxford: Oxford University Press.

Gell, A. (1999) *The Art of Anthropology: Essays and Diagrams*, London: Athlone.

Gendzier, I. (1997) *Notes from the Minefield: United States Intervention in Lebanon and the Middle East, 1945–1958*, New York: Columbia University Press.

Ghambashidze, D. (1919) *Mineral Resources of Georgia and Caucasia: Manganese Industry of Georgia*, London: Allen & Unwin.

Ghazvinian, J. (2007) *Untapped: The Scramble for Africa's Oil*, Orlando, FA: Harcourt.

Gibson-Graham, J.K. (1996) *The End of Capitalism (As We Knew It): A Feminist Critique of Political Economy*, Cambridge, MA: Blackwell.

Gibson-Graham, J.K. (2006) *A Postcapitalist Politics*, Minneapolis, MN: Minnesota University Press.

Gibson-Graham J.K., Resnick, S. and Wolff, R.D. (eds) (2001), *Re/presenting Class: Essays in Postmodern Marxism*, Durham NC: Duke University Press.

Gilmartin, M. (2009) 'Border Thinking: Rossport, Shell and the Political Geographies of a Gas Pipeline', *Political Geography*, 28, 274–282.

Gizzatov, V. (2004) 'Negotiations on the Legal Status of the Caspian Sea 1992–1996: View from Kazakhstan', in S. Akiner (ed.) *The Caspian: Politics, Energy, Security*, London: RoutledgeCurzon, 48–59.

Gökay, B. (ed.) (1999) *The Politics of Caspian Oil*, Basingstoke: Palgrave.

Gordadze, T. (2009) 'Georgian-Russian Relations in the 1990s', in S. Cornell and F. Starr (eds) (2009) *The Guns of August 2008: Russia's War with Georgia*, Armonk, NY: M.E. Sharpe, 28–48.

Gordon, A.G. (2001) *Shell, Greenpeace and the Brent Spar*, Basingstoke: Palgrave.

Gordon, A.G. (2002) 'Indirect Causes and Effects in Policy Change: The Brent Spar Case', *Public Administration*, 76, 4, 713–740.

Gorman, D.G. and Neilson, J. (eds) (1997) *Decommisioning Offshore Structures*, Berlin: Springer-Verlag.

Goudie, A. (2006) *The Human Impact on the Natural Environment*, Oxford: Basil Blackwell.

Gouldson, A. and Bebbington, J., (2007) 'Corporations and the Governance of Environmental Risk', *Environment and Planning C: Government and Policy*, 25, 1, 4–20.

Graham, S. and Thrift, N. (2007) 'Out of Order – Understanding Repair and Maintenance', *Theory Culture & Society*, 24, 3, 1–25.

Gramling, R. and Freudenberg, W. (1992) 'The Exxon Valdez in the Context of US Petroleum Politics', *Organization and Environment*, 6, 3, 175–196.

Gramsci, A. (1971) *Selections from Prison Notebooks*, London: Lawrence and Wishart.

Gramsci, A. (1994) 'The Revolution against Capital', in *Pre-Prison Writings*, Cambridge: Cambridge University Press, 39–42.

Granmayeh, A. (2004) 'Legal History of the Caspian Sea', in S. Akiner (ed.) *The Caspian: Politics, Energy, Security*, London: Routledge Curzon, 17–47.

Grant, B. and Yalçin-Heckmann, L. (eds) (2007) *Caucasus Paradigms: Anthropologies, Histories, and the Making of World Area*, Munster: Lit Verlag.

Greco, M. (2005) 'On the Vitality of Vitalism', *Theory, Culture, and Society*, 22, 1, 15–27.

Green, S. (2005) *Notes from the Balkans: Locating Marginality and Ambiguity on the Greek-Albanian Border*, Princeton, NJ: Princeton University Press.

Greenhough, B. (2006) 'Imagining an Island Laboratory: Representing the Field in Geography and Science Studies', *Transactions of the Institute of British Geographers*, 31, 2, 224–237.

Gregory, D. (2004) *The Colonial Present: Afghanistan, Palestine, Iraq*, Oxford: Basil Blackwell.

Gregory, D. and Pred, A. (eds) (2007) *Violent Geographies: Fear, Terror, and Political Violence*, London: Routledge.

Gregson, N. and Crang, M. (2010) 'Materiality and Waste: Inorganic Vitality in a Networked World', *Environment and Planning A*, 42, 1026–1032.

Gregson, N., Watkins, H. and Calestani, M. (2013) 'Political Markets: Recycling, Economization and Marketization', *Economy & Society*, 42, 1–25.

Gregson, N., Crang, M., Ahamed, F., Akter, N. and Ferdous, R. (2010) 'Following Things of Rubbish Value: End-of-life Ships, "Chock-chocky" Furniture and the Bangladeshi Middle Class Consumer', *Geoforum*, 41, 846–854.

Gross, M. (2010) *Ignorance and Surprise: Science, Society and Ecological Design*, Cambridge, MA: MIT Press.

Grossman, E., Emilio, L. and Muniesa, F. (2008) 'Economies Through Transparency', in C. Garsten and G. de Montoya (eds) (2008) *Transparency in a New Global Order: Unveiling Organizational Visions*, Cheltenham: Edward Elgar, 97–12.

Guliyev, F. and Akhrarkhodjaeva, N. (2009) 'The Trans-Caspian Energy Route: Cronyism, Competition and Cooperation in Kazakh Oil Export', *Energy Policy* 37, 8, 3171–3182.

Gupta, A. (1995) 'Blurred Boundaries: The Discourse of Corruption, the Culture of Politics, and the Imagined State', *American Ethnologist*, 22, 2, 375–402.

Gupta, A. (2008) 'Transparency Under Scrutiny: Information Disclosure in Global Environmental Governance', *Global Environmental Politics*, 8, 2, 1–7.

Guthrie, R. (2002) 'The Effects of Logging on Frequency and Distribution of Landslides in Three Watersheds on Vancouver Island, British Columbia', *Geomorphology*, 43, 3–4, 273–292.

Haber, S. and Menaldo, V. (2011) 'Do Natural Resources Fuel the Resource Authoritarianism? A Reappraisal of the Resource Curse, *American Political Science Review*, 105, 1, 1–26.

Habermas, J. (1971) 'The Scientization of Politics and Public Opinion', in *Toward a Rational Society*, Boston, MA: Beacon Press, 62–80.

Habermas, J. (1990) *The Structural Transformation of the Public Sphere*, Cambridge: Polity.

Hajer, M. (2003) 'Policy without Polity: Policy Analysis and the Institutional Void' *Policy Sciences*, 36, 2, 175–195.

Hajer, M. (2009) *Authoritative Governance: Policy-Making in the Age of Mediatization*, Oxford: Oxford University Press.

Halewood, M. (2011) *A.N. Whitehead and Social Theory: Tracing a Culture of Thought*, London: Anthem Press.

Hamilton, K. (2004) 'Producing Civil Society: Development Intervention and its Consequences in Post-Soviet Georgia', unpublished DPhil thesis, University of Sussex.

Harremoës, P., Gee, D., MacGarvin, M., Stirling, A., Keys, J., Wynne, B. and Guedes Vaz, S. (eds) (2002) *The Precautionary Principle: Late Lessons from Early Warnings*, London: Routledge.

Harvey, D. (2005) *A Brief History of Neo-Liberalism*, Oxford: Oxford University Press.

Harvey, D. (2006) *The Limits to Capital*, London: Verso.

Harvey, P. (2010) 'Cementing Relations: The Materiality of Roads and Public Spaces in Provincial Peru', *Social Analysis*, 54, 2, 28–46.

Harvey, P. and Knox, H. (2010) 'Abstraction, Materiality and the "Science and the Concrete" in Engineering Practice', in T. Bennett and P. Joyce (eds) *Material Powers: Cultural Studies, History and the Material Turn*, London: Routledge, 124–142.

Hawkins, H. (2013) 'Geography and Art. An Expanding Field: Site, the Body and Practice', *Progress in Human Geography*, 37, 1, 52–71.

Hayden, C. (2010) 'The Proper Copy: The Insides and Outsides of Domains Made Public', *Journal of Cultural Economy*, 3, 1, 85–102.

Held, D. and Koenig-Archibugi, M. (eds) (2005) *Global Governance and Public Accountability*, Oxford: Basil Blackwell.

Herzig, E. (1999) *The New Caucasus: Armenia, Azerbaijan and Georgia*, London: RIIA.

Hetherington, K. (2011) *Guerilla Auditors: The Politics of Transparency in Neoliberal Paraguay*, Durham, NC: Duke University Press.

Heyat, F. (2002) *Azeri Women in Transition: Women in Soviet and Post-Soviet Azerbaijan*, London: RoutledgeCurzon.

Hibou, B. (ed.) (2004) *Privatising the State*, London: Hurst.

Hibou, B. (2011) *The Force of Obedience: The Political Economy of Repression in Tunisia*, Cambridge: Polity.

Hinchliffe, S. (2001) 'Indeterminacy in-decisions – Science, Policy and Politics in the BSE Crisis', *Transactions of the Institute of British Geographers*, 26, 2, 182–204.

Hinchliffe, S. (2007) *Geographies of Nature: Societies, Environments, Ecologies*, London: Sage.

Hinchliffe, S. and Bingham, N. (2008) 'Securing Life: The Emerging Practices of Biosecurity. *Environment and Planning A*, 40, 7, 1534–1551.

Hoffman, D. (2000) 'Azerbaijan: The Politicization of Oil', in R. Ebel and R. Menon (eds) *Energy and Conflict in Central Asia and the Caucasus*, New York: Rowman and Littlefield.

Holzer, B. (2010) *Moralizing the Corporation: Transnational Activism and Corporate Accountability*, Cheltenham: Edward Elgar.

Honig, B. (1993) *Political Theory and the Displacement of Politics*, Ithaca, NY: Cornell University Press.

Hood, C. (2006) 'Transparency in Historical Perspective', *Proceedings of the British Academy*, 135, 3–23.

Hood, C. and Heald, D. (eds) (2006) *Transparency – The Key to Better Governance?* Oxford: OUP/British Academy.

Humphrey, C. (2002) *Unmaking of Soviet Life: Everyday Economies after Socialism*, Ithaca, NY: Cornell University Press.

Humphrey, C. (2003) 'Rethinking Infrastructure: Siberian Cities and the Great Freeze of January 2001', in J. Schneider and I. Susser (eds) *Wounded Cities: Destruction and Reconstruction in a Globalized World*, Oxford: Berg.

Humphreys, M. (2005) 'Natural Resources, Conflict and Conflict Resolution', *Journal of Conflict Resolution*, 49, 508–537.

Humphreys, M., Sachs, J. and Stiglitz, J. (2007) *Escaping the Resource Curse*, New York: Columbia University Press.

Huxham, M. and Sumner, D. (1999) 'Emotion, Science and Rationality: The Case of Brent Spar', *Environmental Values*, 8, 349–368.

Jasanoff, S. (1997) 'Civilisation and Madness: The Great BSE Scare of 1996', *Public Understanding of Science*, 6, 3, 221–232.

Jasanoff, S. (ed.) (2006a) *States of Knowledge: The Co-production of Science and Social Order*, London: Routledge.

Jasanoff, S. (2006b) 'Transparency in Public Science: Purposes, Reasons, Limits', *Law and Contemporary Problems*, 69, 21–45.

Jeffrey, A. (2013) *The Improvised State: Sovereignty, Performance and Agency in Dayton Bosnia*, Oxford: Wiley-Blackwell.

Jensen, N. and Johnston, N. (2011) 'Political Reputation, Risk, and the Resource Curse', *Comparative Political Studies*, 44, 6, 662–688.

Jessop, B. (2008) *State Power*, Cambridge: Polity.

Jeter, J. (2006) 'Environmental Due Diligence for International Finance of Major Oil and Gas Projects', SPE 108908, paper presented at the SPE conference on Health, Safety, and Environment in Oil and Gas Exploration and Production held in Abu Dhabi, UAR, 2–4th April.

Johnston, R. and Pattie, C. (2006) *Putting Voters in their Place: Geography and Elections in Great Britain*, Oxford: Oxford University Press.

Jordan, G. (2001) *Shell, Greenpeace and Brent Spar*, Basingstoke: Palgrave Macmillan.

Kaldor, M. (2007) 'Oil and Conflict: The case of Nagorno Karabakh', in M. Kaldor, T.L. Karl and Y. Said (eds) *Oil Wars*, London: Pluto, 157–182.

Kalyuzhnova, Y. (2008) *Economics of the Caspian Oil and Gas Wealth: Companies, Governments, Policies*, Basingstoke: Palgrave Macmillan.

Karl, T.L. (1997) *The Paradox of Plenty: Oil Booms and Petro-States*, Berkeley, CA: California University Press.

Kautsky, K. (1921) *Georgia: A Social-Democratic Peasant Republic – Impressions and Observations*, London: International Bookshops.

Kearns, G. (2004) 'The Political Pivot of History', *The Geographical Journal*, 170, 4, 337–346.

Kearns, G. (2009) *Geopolitics and Empire: The Legacy of Halford Mackinder*, New York: Oxford University Press.

Kellogg, P. (2003) 'The Geo-economics of the New Great Game', *Contemporary Politics*, 9, 1, 75–82.

Kennedy, J. (1993) *Oil and Gas Pipeline Fundamentals*, 2nd edition, Tulsa, OK: PennWell.

Kleveman, L. (2003) *The New Great Game: Blood and Oil in Central Asia*, London: Atlantic Books.

Köksal, S. (2004) 'Oil Pipelines in the CIS, Azeri-Turkish Relations', unpublished PhD thesis, University of London.

Kropp, C. (2005) 'River Landscaping in Second Modernity', in B. Latour and P. Weibel (eds) *Making things Public: Atmospheres of Democracy*, Cambridge, MA: MIT Press, 486–491.

Kukhianidze, A. (2009) 'Corruption and Organized Crime in Georgia Before and After the "Rose Revolution"', *Central Asian Survey*, 28, 2, 215–234.

Kwinter, S. (2001) *Architectures of Time: Towards a Theory of the Event in Modernist Culture*, Cambridge, MA: MIT Press.

Laclau, E. (2005) *Populist Reason*, London: Verso.

Landström, C., Whatmore, S.J., Lane, S.N., Odoni, N.A., Ward, N. and Bradley, S. (2011), 'Coproducing Flood Risk Knowledge: Redistributing Expertise in Critical "Participatory Modelling"' *Environment and Planning A* 43, 7, 1617–1633.

Larner, W. (2000) 'Neo-liberalism: Policy, Ideology, Governmentality', *Studies in Political Economy*, 63, 5–25.

Larner, W. (2009) 'Neoliberalism, Mike Moore and the WTO', *Environment and Planning A*, 41, 7, 1576–1593.

Larner, W. and Walters, W. (eds) (2004) *Global Governmentality*, London: Routledge.

Latham, J. (1986) *Report of a Surveyor*, London: Tate Gallery, quoted in H. Slater (1999) 'The Art of Governance – On The Artists Placement Group 1966–1989', http://www.variant.org.uk/pdfs/issue11/Howard_Slater.pdf (accessed 16 May 2013).

Latour, B. (1987) *Science in Action*, Milton Keynes: Open University Press.

Latour, B. (1999) *Pandora's Hope: Essays on the Reality of Science Studies*, Cambridge, MA: Harvard University Press.

Latour, B. (2004) *The Politics of Nature*, Cambridge, MA: Harvard University Press.

Latour, B. (2005a) 'From *Realpolitik* to *Dingpolitik*: Or How to Make Things Public', in B. Latour and P. Weibel (eds) *Making things Public: Atmospheres of Democracy*, Cambridge, MA: MIT Press, 14–41.

Latour, B. (2005b) *Reassembling the Social: An Introduction to Actor-Network Theory*, Oxford: Oxford University Press.

Latour, B. (2009) *The Making of Law: An Ethnography of the Conseil d'État*, Cambridge: Polity.

Latour, B. and Weibel, P. (eds) (2005) *Making things Public: Atmospheres of Democracy*, Cambridge, MA: MIT Press.

Latour, B. and Woolgar, S. (1986) *Laboratory Life: The Construction of Scientific Facts*, Princeton, NJ: Princeton University Press.

Laurent, B. (2011) 'Democracies on Trial: Assembling Nanotechnology and its Problems', unpublished doctoral thesis, Mines ParisTech.

Law, J. (2002) *Aircraft Stories: Decentring the Object in Technoscience*, Durham, NC: Duke University Press.

Law, J. (2004) *After Method: Mess in Social Research*, London: Routledge.

Law, J. and Mol, A. (2008) 'Globalisation in Practice: The Politics of Boiling Pigswill', *Geoforum*, 39, 1, 133–143.

Law, J. and Urry, J. (2005) 'Enacting the Social', *Economy and Society*, 33, 3, 390–410.

Lawrence, R. (2009) 'Shifting Responsibilities and Shifting Terrains: State Responsibility, Corporate Responsibility and Indigenous Claims', unpublished PhD thesis, Department of Sociology, Stockholm University.

Lazar, S. (2008) *El Alto, Rebel City: Self and Citizenship in Andean Bolivia*, Durham, NC: Duke University Press.

Lazzarato, M. (1996) 'Immaterial Labour', in P. Virno and M. Hardt (eds) *Radical Thought in Italy: A Potential Politics*, Minneapolis, MN: Minnesota University Press.

Le Billon, P. (ed.) (2005) *The Geopolitics of Resource Wars: Resource Dependence Governance and Violence*, London: Routledge.

Ledeneva, A. (1998) *Russia's Economy of Favours: Blat, Networking and Informal Exchange*, Cambridge: Cambridge University Press.

Lee, E.M. (2009) 'Landslide Risk Assessment: The Challenge of Estimating the Probability of Landsliding', *Quarterly Journal of Engineering Geology and Hydrogeology*, 42, 4, 445–458.

Lee, E.M. and Charman, J.H. (2005) 'Geohazards and Risk Assessment for Pipeline Route Selection', in M. Sweeney (ed.) *Terrain and Geohazard Challenges Facing Onshore Oil and Gas Pipelines*, London: Thomas Telford, 95–116.

Lee, E.M. and Jones, D. (2004) *Landslide Risk Assessment*, London: Thomas Telford.

Leech, G. (2006) *Crude Interventions: The United States, Oil and the New World Disorder*, London: Zed.

Lefebvre, H. (1984) *The Production of Space*, Oxford: Basil Blackwell.

Lemos, M. and Agrawal, A. (2006), 'Environmental Governance', *Annual Review of Environment and Resources*, 31, 3, 297–325.

Lerman, Z. (2006) 'The Impact of Land Reform on Rural Household Incomes in Transcaucasia', *Eurasian Geography and Economics*, 47, 1, 112–123.

Levidow, L. (2001). 'Precautionary Uncertainty: Regulating GM Crops in Europe. *Social Studies of Science*, 31, 6, 842–874.

Lezaun, J. and Soneryd, L. (2007) 'Consulting Citizens: Technologies of Elicitation and the Mobility of Politics', *Public Understanding of Science*, 16, 279–297.

Li, T.M. (2007) *The Will to Improve: Governmentality, Development and the Practice of Politics*, Durham, NC: Duke University Press.

Lippard, L. (1973) *Six Years: The Dematerialisation of the Art Object from 1966 to 1972*, Berkeley, CA: University of California Press.

Livesey, S. (2001) 'Eco-Identity as Discursive Struggle: Royal Dutch/Shell, Brent Spar and Nigeria', *Journal of Business Communication*, 38, 1, 58–91.

Livingstone, D. (1992) *The Geographical Tradition: Episodes in a History of a Contested Enterprise*, Oxford: Basil Blackwell.

Livingstone, D. (2003) *Putting Science in its Place: Geographies of Scientific Knowledge*, Chicago, IL: Chicago University Press.

Lloyd Thomas, K. (2010) 'Building Materials: Conceptualising Materials via the Architectural Specification', unpublished PhD thesis, Department of Philosophy, Middlesex University.

Löfstedt, R. and Renn, O. (1997) 'The Brent Spar Controversy: An Example of Risk Communication Gone Wrong', *Risk Analysis*, 17, 2, 131–136.

Luhmann, N. (2002) *Risk: A Sociological Theory*, Brunswick, NJ: Transaction.

Lury, C. (2004) *Brands: The Logos of the Global Economy*, London: Routledge.

Lussac, S. (2010a) *Géopolitque du Caucase: au carrefour énergétique de l'Europe de l'Ouest*, Paris: Éditions Technip.

Lussac, S. (2010b) 'The State as an (Oil) Company? The Political Economy of Azerbaijan', GARNET working paper 74/10, Sciences Po, Bordeaux.

Lynch, M. (1998) 'The Discursive Production of Uncertainty: The O.J. Simpson "Dream Team" and the Sociology of Knowledge Machine', *Social Studies of Science*, 28, 5–6, 829–868.

Maas, P. (2009) *Crude World: The Violent Twilight of Oil*, London: Penguin.

Macdonald, F., Hughes, R. and Dodds, K. (eds) (2010) *Geopolitics and Visual Culture*, London: I.B. Tauris.

MacIntosh, N. and Quattrone, P. (2010) *Management Accounting and Control Systems: An Organisational and Sociological Approach*, New York: Wiley.

MacIntyre, A. (1981) *After Virtue: A Study in Moral Theory*, London: Duckworth.

MacKenzie, A. (2002) *Transductions: Bodies and Machines at Speed*, London: Continuum.

MacKenzie, D. (1996) *Knowing Machines: Essays on Technical Change*, Cambridge, MA: MIT Press.

MacKenzie, D., Muniesa, F. and Siu, L. (eds) (2007) *Do Economists Make Markets? On the Performativity of Economics*, Princeton, NJ: Princeton University Press.

Mackinder, H. (1887) 'The Scope and Methods of Geography', *Proceedings of the Royal Geographical Society*, 9, in H. Mackinder (1962) *Democratic Ideals and Reality*, New York: W.W. Norton, 211–240.

Mackinder, H. (1904) 'The Geographical Pivot of History', *The Geographical Journal*, XXIII, in H. Mackinder (1962) *Democratic Ideals and Reality*, New York: W.W. Norton, 241–264.

Mackinder, H. (1962) *Democratic Ideals and Reality*, New York: W.W. Norton.

MacLean, F. (1949) *Eastern Approaches*, London: Jonathan Cape.

Manning, P. (2007) 'Rose-colored Glasses? Color Revolutions and Cartoon Chaos in Post-socialist Georgia', *Cultural Anthropology*, 22, 2, 171–213.

Marres, N. (2005) 'No Issue, No Publics: Democratic Deficits and the Displacement of Politics', unpublished PhD thesis University of Amsterdam.

Marres, N. (2012) *Material Participation: Technology, the Environment and Everyday Publics*, Basingstoke: Palgrave Macmillan.

Marres, N. and Rogers, R. (2008) 'Subsuming the Ground: How Local Realities of the Fergana Valley, the Narmada Dams and the BTC Pipeline Are Put to Use on the Web', *Economy and Society*, 37, 2, 251–281.

Marx, K. (1973 [1852]) *The Eighteenth Brumaire of Louis Bonaparte, in Surveys from Exile*, Harmondsworth: Penguin.

Marx, K. (1976 [1867]) *Capital, volume 1*, Harmondsworth: Penguin.

Massey, D. (2005) *For Space*, London: Sage.

Massey, D. (2010) *World City*, Cambridge: Polity.

May, J. and Thrift, N. (eds) (2001) *Timespace: Geographies of Modernity*, London: Routledge.

McCormack, D. (2007) 'Molecular Affects in Human Geographies'. *Environment and Planning A*, 39, 2, 359–377.

McGoey, L. (2007) 'On the Will to Ignorance in Bureaucracy', *Economy and Society*, 36, 2, 212–235.

McGoey, L. (2009) 'Pharmaceutical Controversies and the Performative Value of Uncertainty', *Science as Culture*, 18, 2, 151–164.

McNeish, J.-A. and Logan, O. (eds) (2012) *Flammable Societies: Studies in the Socio-economics of Oil and Gas*, London: Pluto.

Megoran, N. (2006) 'For Ethnography in Political Geography: Experiencing and Re-imagining Ferghana Valley Boundary Closures', *Political Geography*, 25, 622–640.

Mitchell, L. (2009) *Uncertain Democracy: US Foreign Policy and Georgia's Rose Revolution*, Philadelphia, PA: University of Pennsylvania Press.

Mitchell, T. (1999) 'Society, Economy and the State Effect', in G. Steinmetz (ed.) *State/Culture: State-Formation after the Cultural Turn*, Durham, NC: Duke University Press, 76–97.

Mitchell, T. (ed) (2000) *Questions of Modernity*, Minneapolis, MN: Minnesota University Press.

Mitchell, T. (2002) *Rule of Experts: Egypt, Techno-Politics, Modernity*, Berkeley, CA: California University Press.

Mitchell, T. (2005) 'The Work of Economics: How a Discipline Makes its World', *European Journal of Sociology*, 46, 2, 297–320.

Mitchell, T. (2007) 'The Properties of Markets', in D. MacKenzie, F. Muniesa and L. Siu (eds) *Do Economists Make Markets?: On the Performativity of Economics*, Princeton, NJ: Princeton University Press, 244–275.

Mitchell, T. (2008) 'Rethinking Economy', *Geoforum*, 39, 1116–1121.

Mitchell, T. (2011) *Carbon Democracy: Political Power in the Age of Oil*, London: Verso.

Mol, A. (2002) *The Body Multiple: Ontology in Medical Practice*, Durham, NC: Duke University Press.

Mol, A. (2006) *The Logic of Care: Health and the Problem of Patient Choice*, London: Routledge.

Monk, R. (1996) *Bertrand Russell: The Spirit of Solitude*, London: Jonathan Cape.

Morris, P. and Therivel, R. (eds) (2009) *Environmental Impact Assessment*, 3rd edition, London: Routledge.

Mouffe, C. (1993) *The Return of the Political*, London: Verso.

Mouffe, C. (2000) *The Democratic Paradox*, London: Verso.

Mouffe, C. (2005a) *On the Political*, London: Routledge.

Mouffe, C. (2005b) 'Some Reflections on an Agonistic Approach to the Public' in B. Latour and P. Weibel (eds) *Making things Public: Atmospheres of Democracy*, Cambridge, MA: MIT Press, 804–809.

Muniesa, F. and Callon, M. (2007) 'Economic Experiments and the Construction of Markets", in D. MacKenzie, F. Muniesa and L. Siu (eds) *Do Economists Make Markets? On the Performativity of Economics*, Princeton, NJ: Princeton University Press, 163–189.

Muniesa, F., Millo,Y. and Callon, M. (2007) 'An Introduction to Market Devices', in M. Callon,Y. Millo and F. Muniesa (eds) *Market Devices*, Oxford: Basil Blackwell.

Murdoch, J. (2006) *Post-structuralist Geography*, London: Sage.

Muttit, G. (2011) *Fuel on the Fire: Oil and Politics in Occupied Iraq*, London: Bodley Head.

Nancy, J.-L. (2007) *Listening*, New York: Fordham University Press.

Navaro-Yashin, Y. (2002) *Faces of the State: Secularism and Public Life in Turkey*, Princeton, NJ: Princeton University Press.

Navaro-Yashin, Y. (2012) *The Make-Believe Space: Affective Geography in a Post-War Polity*, Durham, NC: Duke University Press.

Neale, S. (1980) *Genre*, London: Bfi.

Newman, D. (2003) 'Boundaries', in J. Agnew, K. Mitchell and G. Toal (eds) *A Companion to Political Geography*, Oxford: Basil Blackwell, 123–137.

Neyland, D. (2007) 'Achieving Transparency: The Visible, Invisible and Divisible in Academic Accountability Networks', *Organization*, 14, 4, 499–516.

Noble, B. (2010) *Introduction to Environmental Impact Assessment*, 3rd edition, Oxford: Oxford University Press.

Nodia, G. and Scholtbach, A.P. (eds) (2006) *The Political Landscape of Georgia*, Delft: Eburon.

Nourzhanov, K. (2006) 'Caspian Oil: Geopolitical Dreams and Real Issues', *Australian Journal of International Affairs*, 60, 1, 59–66.

November, V., Camacho-Hübner, E. and Latour, B. (2010) 'Entering a Risky Territory: Space in the Age of Digital Navigation', *Environment and Planning D: Society and Space*, 28, 4, 581–599.

Nowotny, H., Scott, P. and Gibbons, M. (2001) *Rethinking Science: Knowledge and the Public in an Age of Uncertainty*, Cambridge: Polity.

O'Hara, S. (2005) 'Great Game or Grubby Game?: The Struggle for Control of the Caspian', in P. Le Billon (ed.) *The Geopolitics of Resource Wars: Resource Dependence Governance and Violence*, London: Routledge, 138–160.

Okanta, I. and Douglas, O. (2003) *Where Vultures Feast: Shell, Human Rights, and Oil in the Niger Delta*, London: Verso.

Oldfield, J., O'Hara, S. and Shaw, D. (2004) 'Environment and Environmentalism', in M. Bradshaw and A. Stenning (eds) *East Central Europe and the Former Soviet Union: The Post-Socialist States*, Harlow: Pearson, 137–160.

O'Lear, S. (2005) 'Resources and Conflict in the Caspian Sea', in P. Le Billon (ed.) *The Geopolitics of Resource Wars: Resource Dependence, Governance and Violence*, London: Routledge, 161–186.

O'Lear, S. (2007) 'Azerbaijan's Resource Wealth: Political Legitimacy and Public Opinion', *Geographical Journal*, 173, 3, 207–223.

O'Lear, S. and Gray, A. (2006) 'Asking the Right Questions: Environmental Conflict in the Case of Azerbaijan', *Area*, 38, 4, 390–401.

Ong, A. (2007) 'Neoliberalism as a Mobile Technology', *Transactions of the Institute of British Geographers*, 32, 1, 3–8.

Ong, A. and Collier, S. (eds) (2005) *Global Assemblages: Technology, Politics and Ethics as Anthropological Problems*, Oxford: Basil Blackwell.

Osborne, T. (1993) 'On Liberalism, Neo-Liberalism and the "Liberal Professions" of Medicine', *Economy and Society*, 22, 3, 345–356.

Osborne, T. (1998) *Aspects of Enlightenment: Social Theory and the Ethics of Truth*, London: UCL Press.

Osborne, T. (1999) 'Critical Spirituality: Ethics and Politics in the Later Foucault', in S. Ashenden and D. Owen (eds) *Foucault contra Habermas*, London: Sage.

Osborne, T. (2004) 'On Mediators: On Intellectuals and the Ideas Trade in the Knowledge Society', *Economy and Society*, 33, 4, 430–447.

Osborne, T. and Rose, N. (1999) 'Do the Social Sciences Create Phenomena?: The Case of Public Opinion Research', *British Journal of Sociology*, 50, 3, 367–396.

O'Tuathail, G. (1996) *Critical Geopolitics: The Politics of Writing Global Space*, Minneapolis, MN: University of Minnesota Press.

O'Tuathail, G. and Dalby, S. (eds) (1998) *Rethinking Geopolitics*, London: Routledge.

O'Tuathail, G., Dalby, S. and Routledge, P. (eds) (1998) *The Geopolitics Reader*, London: Routledge.

Overland, I., Kjaernet, H. and Kendall-Taylor, A. (eds) (2010) *Caspian Energy Politics: Azerbaijan, Kazakhstan and Turkmenistan*, London: Routledge.

Peel, M. (2009) *A Swamp Full of Dollars: Pipelines and Paramilitaries at Nigeria's Oil Frontier*, London: I.B. Tauris.

Peirce, C.S. (1934) 'Pragmatism and Abduction', in C. Hartshorne and P. Weiss (eds) *Collected Papers of Charles Sanders Peirce vol 5*, Cambridge, MA: Harvard University Press.

Petryna, A. (2002) *Life Exposed: Biological Citizens after Chernobyl*, Princeton, NJ: Princeton University Press.

Pickering, A. (1981) 'Constraints on Controversy: The Case of the Magnetic Monopole', *Social Studies of Science*, 11, 63–93.

Pickering, A. (1995) *The Mangle of Practice: Time, Agency and Science*, Chicago, IL: Chicago University Press.

Pollett, E. and Wyness, R. (2006) 'Social Impact Assessment and Mitigation for a Large Pipeline Project', SPE 98861, paper presented at the SPE conference on Health, Safety, and Environment in Oil and Gas Exploration and Production held in Abu Dhabi, UAR, 2–4th April.

Powell, R. (2007) 'Geographies of Science: Histories, Localities, Practices, Futures', *Progress in Human Geography*, 31, 3, 309–329.

Powell, R. (2010) 'Lines of Possession? The Anxious Constitution of a Polar Geopolitics', *Political Geography*, 29, 2, 74–77.

Power, M. (1997) *The Audit Society: Rituals of Verification*, Oxford: Oxford University Press.

Power, M. (2007a) *Organised Uncertainty: Designing a World of Risk Management*, Oxford: Oxford University Press.

Power, M. (2007b) 'Corporate Governance, Reputation and Environmental Risk', *Environment and Planning C*, 25, 1, 90–97.

Power, M. and Campbell, D. (2010) 'The State of Critical Geopolitics', *Political Geography*, 29, 5, 243–246.

Quattrone, P. (2006) 'The Possibility of the Testimony: The Case of Case Study Research', *Organisation*, 13, 1, 143–157.

Rajak, D. (2011) *In Good Company: An Anatomy of Corporate Social Responsibility*, Stanford, CA: Stanford University Press.

Rancière, J. (1998) *Disagreement: Politics and Philosophy*, Minneapolis, MN: Minnesota University Press.

Rancière, J. (2001) 'Ten Theses on Politics', *Theory and Event*, 5, 3.

Rancière, J. (2004a) *The Politics of Aesthetics*, London: Continuum.

Rancière, J. (2004b) *The Philosopher and his Poor*, Durham, NC: Duke University Press.

Rancière, J. (2006) *Hatred of Democracy*, London: Verso.

Rasizade, A. (2002) 'The Mythology of Munificent Caspian Bonanza and its Concomitant Pipeline Politics', *Central Asian Survey*, 21, 1, 37–54.

Redgewell, C. (2012) 'Contractual and Treaty Arrangements Supporting Large European Transboundary Pipeline Projects: Can Adequate Human Rights and Environmental Protection be Secured?', in M. Roggenkamp, L. Barrera-Hernandez, D. Zillman, and I. del Guayo (eds) *Energy Networks in the Law: Innovative Solutions in Changing Markets*, Oxford: Oxford University Press, 102–117.

Reed, K. (2009) *Crude Existence: Environment and the Politics of Oil in Northern Angola*, Berkeley, CA: University of California Press.

Reeves, M. (2008) 'Border Work: An Ethnography of the State and its Limits', unpublished PhD thesis, Department of Social Anthropology, University of Cambridge.

Reeves, M. (2011) 'Fixing the Border: On the Affective Life of the State in Southern Kyrgyzstan', *Environment and Planning D: Society and Space*, 29, 5, 905–923.

Regester, M. and Larkin, J. (2005) *Risk Issues and Crisis Management: A Casebook of Best Practice* 3rd edition, London: Kogan Page/Chartered Institute of Public Relations.

Rice, T. and Owen, P. (1999) *Decommissioning the Brent Spar*, London: E and FN Spon/Routledge.

Riles, A. (2001) *The Network Inside Out*, Ann Arbor, MI: Michigan University Press.

Riles, A. (2006) '[Deadlines]: Removing the Brackets on Politics in Bureaucratic and Anthropological Analysis', in A. Riles (ed), *Documents*, Ann Arbor, MI: University of Michigan Press, 71–94.

Roberts, J. (1996) *Caspian Pipelines*, London: RIIA.

Roberts, J. (2004) 'Pipeline Politics' in S. Akiner (ed.) *The Caspian: Politics, Energy, Security*, London: Routledge Curzon, 77–89.

Robins, P. (2002) *Suits and Uniforms: Turkish Foreign Policy and the Cold War*, London: Hurst.

Rogers, D. (2012) 'The Materiality of the Corporation: Oil, Gas, and Corporate Social Technologies in the Remaking of a Russian Region', *American Ethnologist*, 39, 2, 284–296.

Rogers, R. (2004) *Information Politics on the Web*, Cambridge, MA: MIT Press.

Rose, C. (1998) *The Turning of the Spar*, London: Greenpeace.

Rosengarten, M. (2009) *HIV Interventions: Biomedicine and the Traffic between Information and Flesh*, Seattle, WA: University of Washington Press.

Ross, M. (2001) 'Does Oil Hinder Democracy?', *World Politics*, 53, 325–361.

Rosser, A. (2006) 'Escaping the Resource Curse', *New Political Economy*, 11, 4, 557–570.

Roux, J. and Magnin, T. (2004), *La condition de fragilité: entre science des matériaux et sociologie*, St Étienne: Publications de l'Université de Saint-Étienne.

Rowell, A., Marriott, J. and Stockman, L. (2005) *The Next Gulf : London, Washington and Oil Conflict in Nigeria*, London: Robinson.

Runciman, D. (2006) *The Politics of Good Intentions: History, Fear and Hypocrisy in the New World Order*, Princeton, NJ: Princeton University Press.

Russell, B. (1920) *The Practice and Theory of Bolshevism*, London: George Allen and Unwin.

Russell, B. (1923) *The Prospects of Industrial Civilisation*, London: George Allen and Unwin.

Ryan, A. (1988) *Bertrand Russell: A Political Life*, Harmondsworth: Penguin.

Sachs, J. (1996) 'The Transition at Mid-Decade', *American Economic Review*, 86, 2, 128–133.

Sachs, J. and Warner, A. (2001) 'The Curse of Natural Resources', *European Economic Review*, 45, 827–838.

Sadler, D. (2004) 'Trade Unions, Coalitions and Communities: Australia's Construction, Forestry, Mining and Energy Union and the International Stakeholder Campaign Against Rio Tinto' *Geoforum*. 35, 1, 35–46.

Salmon, L. (2005) 'Gabriel Tarde and the Dreyfus Affair: Reflections on the Engagement of an Intellectual', *Champ pénal: nouvelle revue internationale de criminologie*, vol. II., http://champpenal.revues.org/7185 (last accessed May 2013).

Sawyer, S. (2004) *Crude Chronicles: Indigenous Politics, Multinational Oil, and Neoliberalism in Ecuador*, Durham, NC: Duke University Press.

Schaffer, S. (2003) 'Enlightenment Brought Down to Earth', *History of Science*, 41, 257–268.

Schaffer, S. (2005) 'Public Experiments' in B. Latour and P. Weibel (eds) *Making things Public: Atmospheres of Democracy*, Cambridge, MA: MIT Press, 298–307.

Schmitt, C. (1996 [1932]) *The Concept of the Political*, Chicago, IL: Chicago University Press.

Schueth, S. (2012) 'Apparatus of Capture: Fiscal State Formation in the Republic of Georgia', *Political Geography*, 31, 3, 133–143.

Schultz, M., Hatch, M. and Larsen, M.H. (eds) (2000) *The Expressive Organisation: Linking Identity, Reputation and the Corporate Bond*, Oxford: Oxford University Press.

Schulz, J. (2005) *Follow the Money: A Guide to Monitoring Budgets and Oil and Gas Revenues*, New York: Open Society Institute.

Selley, R. (1998) *Elements of Petroleum Geology*, San Diego, CA: Academic Press.

Shapin, S. and Schaffer, S. (1985) *Leviathan and the Air-Pump: Hobbes, Boyle and the Experimental Life*, Princeton, NJ: Princeton University Press.

Sheail, J. (2007) 'Torrey Canyon: The Political Dimension', *Journal of Contemporary History*, 42, 3, 485–504.

Simmel, G. (1950) 'The Secret and the Secret Society', in K. Wolff (ed.) *The Sociology of Georg Simmel*, Glencoe, IL: The Free Press.

Simondon, G. (1992) 'The Genesis of the Individual', in J. Crary and S. Kwinter (eds) *Incorporations*, New York: Zone.

Slater, H. (1999) 'The Art of Governance – On The Artists Placement Group 1966–1989', http://www.variant.org.uk/pdfs/issue11/Howard_Slater.pdf (accessed 16 May 2013).

Smith, R. (2000) *After the Brent Spar: Business, the Media and the New Environmental Politics*, London: Earthscan.

Soares de Oliveira, R. (2007) *Oil and Politics in the Gulf of Guinea*, London: Hurst.

Ssorin-Chaikov, N. (2003) *The Social Life of the State in Subarctic Siberia*, Stanford, CA: Stanford University Press.

Ssorin-Chaikov, N. (2006) 'On Heterochrony: Birthday Gifts to Stalin, 1949', *Journal of Royal Anthropological Institute*, 12, 355–375.

Stengers, I. (1997) *Power and Invention: Situating Science*, Minneapolis, MN: Minnesota University Press.

Stengers, I. (2000) *The Invention of Modern Science*, Minneapolis, MN: Minnesota University Press.

Stengers, I. and Bensaude-Vincent, B. (2003) *100 Mots pour commencer à penser les Sciences*, Paris: Les Empêcheurs de Penser en Rond.

Stirling, A. (2008) '"Opening Up" and "Closing Down" Power, Participation, and Pluralism in the Social Appraisal of Technology', *Science, Technology and Human Values*, 33, 2, 262–294.

Stoler, A.L. (2009) *Along the Archival Grain: Epistemic Anxieties and Colonial Common Sense*, Princeton, NJ: Princeton University Press.

Strathern, M. (1991) *Partial Connections*, Lanham, MD: Rowman and Littlefield.

Strathern, M. (1995) *The Relation: Issues in Complexity and Scale*, Cambridge: Prickly Pear Press.

Strathern, M. (1999) *Property, Substance and Effect*, London: Athlone.

Strathern, M. (2000) 'The Tyranny of Transparency', *British Educational Research Journal*, 26, 3, 299–321.

Strathern, M. (2004) *Commons and Borderlands: Working Papers on Interdisciplinarity, Accountability and the Flow of Knowledge*, Wantage: Sean Kingston.

Suny, R. (1994) *The Making of the Georgian Nation*, 2nd edition, Bloomington, IN: Indiana University Press.

Swanson, P., Oldgard, M. and Lunde, L. (2003) 'Who Gets the Money? Reporting Resource Revenues', in I. Bannon and P. Collier (eds) *Natural Resources and Violent Conflict*, Washington, DC: The World Bank, 43–96.

Tarde, G. (1969) 'Invention' (extract from Tarde, La Psychologie économique), in *On Communication and Social Influence*, Chicago, IL: Chicago University Press, chapter 6.

Tarde, G. (1999 [1893]) *Monadologie et sociologie*, Paris: Les Empêcheurs de Penser en Rond.

Tarde, G. (2001 [1890]) *Les Lois de l'imitation*, Paris: Les Empêcheurs de Penser en Rond.

Tarde, G. (2006 [1901]) *L'Opinion et la foule*, Paris: Éditions du Sandre.

Taussig, M. (1999) *Defacement: Public Secrecy and the Labor of the Negative*, Stanford, CA: Stanford University Press.

Thoburn, N. (2003) *Deleuze, Marx and Politics*, London: Routledge.

Thompson, G. (2005) 'Global Corporate Citizenship: What Does it Mean?', *Competition and Change*, 9, 2, 131–152.

Thompson, G. (2012) *The Constitutionalization of the Global Corporate Sphere?*, Oxford: Oxford University Press.

Thrift, N. (2000) 'It's the Little Things', in K. Dodds and D. Atkinson (eds) *Geopolitical Traditions: A Century of Geopolitical Thought*, London: Routledge.

Thrift, N. (2005) *Knowing Capitalism*, London: Sage.

Thrift, N. (2006a) 'Space', *Theory, Culture and Society*, 23, 2–3, 139–155.

Thrift, N. (2006b) 'Space, Place, and Time', in R. Goodwin and C. Tilly (eds) *The Oxford Handbook of Contextual Political Analysis*, Oxford: Oxford University, 547–563.

Thrift, N. (2008) *Non-Representational Theory: Space, Politics, Affect*, London: Routledge.

Tilly, C. (1986) *The Contentious French: Four Centuries of Popular Struggle*, Cambridge, MA: Harvard University Press.

Toal, G. and Dahlman, C. (2011) *Bosnia Remade: Ethnic Cleansing and its Reversal*, Oxford: Oxford University Press.

Toscano, A. (2007) 'Powers of Pacification: State and Empire in Gabriel Tarde', *Economy and Society*, 36, 4, 614–643.

Toscano, A. (2008) 'The Culture of Abstraction', *Theory, Culture, and Society*, 25, 4, 57–75.

Trier, T. and Turashvili, M. (2007) *Displacement of Ecologically Displaced Persons: Solution of a Problem or Creation of a New? Eco-migration in Georgia, 1981–2006*, Flensburg: European Centre for Minority Issues.

Trotsky, L. (1975 [1922]) 'Between Red and White: A Study of Some Fundamental Questions of Revolution, With Particular Reference to Georgia', in *Social Democracy and the Wars of Intervention in Russia 1918–21*, London: New Park Publications.

Tsing, A. (2005) *Friction: An Ethnography of Global Connection*, Princeton, NJ: Princeton University Press.

Tsing, A. (2009) 'Supply Chains and the Human Condition', *Rethinking Marxism: A Journal of Economics, Culture & Society*, 21, 2, 148–176.

Urry, J. (1995) *The Tourist Gaze*, London: Sage.

Urry, J. (2003) *Global Complexity*, Cambridge: Polity.

Valdivia, G. (2008) 'Governing Relations between People and Things: Citizenship, Territory, and the Political Economy of Petroleum in Ecuador', *Political Geography*, 27, 456–477.

Venturini, T. (2011) 'Building on Faults: How to Represent Controversies with Digital Methods', *Public Understanding of Science*, 21, 7, 796–812.

Verdery, K. (2003) *The Vanishing Hectare: Property and Value in Postsocialist Transylvania*, Ithaca, NY: Cornell University Press.

Verdery, K. and Humphrey C. (eds) (2004) *Property in Question: Value Transformation in the Global Economy*, Oxford: Berg.

Vitalis, R. (2009) *America's Kingdom: Mythmaking on the Saudi Oil Frontier*, London: Verso.

Waldron, J. (1999a) *The Dignity of Legislation*, Cambridge: Cambridge University Press.

Waldron, J. (1999b) *Law and Disagreement*, Oxford: Oxford University Press.

Walker, R. (2002) *Left Shift: Radical Art in 1970s Britain*, London: I.B. Tauris.

Walker, R.B.J. (1993) *Inside/Outside: International Relations as Political Theory*, Cambridge: Cambridge University Press.

Walker, R.B.J. (2010) *After the Globe, Before the World*, London: Routledge.

Walters, W. (2012) *Governmentality: Critical Encounters*, London: Routledge.

Warner, M. (2002) *Publics and Counter-Publics*, New York: Zone Books.

Watts, M. (2004) 'Resource Curse? Governmentality, Oil and Power in the Niger Delta, Nigeria', *Geopolitics*, 9, 1, 50–80.

Watts, M. (2005) 'Righteous Oil? Human Rights, the Oil Complex and Corporate Social Responsibility', *Annual Review of Environment and Resources*, 30, 9.1–9.35.

Watts, M. (2006) 'Culture, Development and Global Neo-Liberalism', in S. Radcliffe (ed.) *Culture and Development in a Globalizing World: Geographies, Actors, and Paradigms*, London: Routledge, 30–57.

Watts, M. (ed.) (2008) *Curse of the Black Gold: Fifty Years of Oil in the Niger Delta*, New York: powerHouse Books.

West, H. and Sanders, T. (eds) (2003) *Transparency and Conspiracy: Ethnographies of Suspicion in the New World Order*, Durham, NC: Duke University Press.

Weszkalnys, G. (2007) 'Hope and Oil: An Ethnographic Study of the Emergent Oil Operations in São Tomé e Príncipe', report on fieldwork in São Tomé e Príncipe, School of Geography and Environment, Oxford University, http://www.geog.ox.ac.uk/research/technologies/projects/hope-oil.pdf (accessed May 2013).

Weszkalnys, G. (2008) 'Hope and Oil: Expectations in São Tomé e Príncipe', *Review of African Political Economy*, 35, 3, 2008, 473–482.

Weszkalnys, G. (2009) 'The Curse of Oil in the Gulf of Guinea: A view from São Tomé and Príncipe', *African Affairs*, 108, 433, 2009, 679–689.

Weszkalnys, G. (2011) 'Cursed Resources, Or Articulations of Economic Theory in the Gulf of Guinea', *Economy and Society*, 40, 3, 345–372.

Whatmore, S. (2006) 'Materialist Returns: Practicing Cultural Geography In and For a More-than-human World', *Cultural Geographies*, 13, 600–609.

Whatmore, S. (2009) 'Mapping Knowledge Controversies: Science, Democracy and the Redistribution of Expertise', *Progress in Human Geography*, 33, 5, 587–598.

Whatmore, S.J. and Landström, C. (2011) 'Flood Apprentices: An Exercise in Making Things Public', *Economy and Society*, 40, 4, 582–610.

Wheatley, J. (2005) *Georgia from National Awakening to Rose Revolution: Delayed Transition in Post-Soviet Politics*, Aldershot: Ashgate.

Wheatley, J. (2006) 'Defusing Conflict in Tsalka District of Georgia: Migration, International Intervention and the Role of the State', *ECMI working paper 36*, Flensburg: European Centre for Minority Issues.

Wheatley, J. (2009) 'Managing Ethnic Diversity in Georgia: One Step Forward, Two Steps Back', *Central Asian Survey*, 28, 2, 119–134.

Whitehead, A.N. (1920) *The Concept of Nature*, Cambridge: Cambridge University Press.

Whitehead, A.N. (1927) *Symbolism: Its Meaning and Effect*, New York: Macmillan.

Whitehead, A.N. (1933) *The Adventure of Ideas*, Cambridge: Cambridge University Press.

Whitehead, A.N. (1985 [1926]) *Science and the Modern World*, London: Free Association Books.

Wilson, A. (2005) *Virtual Politics: Faking Democracy in the Post-Soviet World*, New Haven, CT: Yale University Press.

Woods, N. (2001) 'Making the IMF and the World Bank More Accountable', *International Affairs*, 77, 1, 83–100.

Woolfson, C., Foster, J. and Beck, M. (1996) *Paying for the Piper: Capital and Labour in Britain's Offshore Oil Industry*, London: Mansell.

Wynne, B. (1996) 'May the Sheep Safely Graze? A Reflexive View of the Expert-Lay Knowledge Divide', in S. Lash, Szerszynski, B. and Wynne, B. (eds) *Risk, Environment and Modernity: Towards a New Ecology*, London: Sage, 44–83.

Yakovleva, N. (2011) 'Oil Pipeline Construction in Eastern Siberia: Implications for Indigenous People', *Geoforum*, 42, 708–719.

Yalçin-Heckmann, L. (2010) *The Return of Private Property: Rural Life after Agrarian Reform in the Republic of Azerbaijan*, Berlin: Lit Verlag.

Yearley, S. (2005) *Cultures of Environmentalism: Empirical Studies in Environmental Sociology*, London: Palgrave Macmillan.

Yergin, D. (1991) *The Prize: The Epic Quest for Oil, Money and Power*, New York: the Free Press.

Zizek, S. (2004) 'Afterword', in J. Rancière, *The Politics of Aesthetics*, London: Continuum.

Film, Television, Video and Photography

Allan, Dominic (2004) *Extreme Oil: The Pipeline* (Thirteen/WNET, US/UK/Canada).

Apted, Michael (1999) *The World is Not Enough* (Eon Productions, UK).

Biemann, Ursula (2005) *Black Sea Files* (Switzerland) http://www.geobodies.org/art-and-videos/black-sea-files (accessed May 2013).

Cran, William (2005) *Storyville – The Curse of Oil* (Paladin InVision/BBC, UK).

Jaar, Alfredo (2006) *Muxima* (Libertoriate).

Kirtadze, Nino (2005) *Un Dragon dans les eaux pures du Caucase* (The Pipeline Next Door ; Roche Productions/Arte, France/Italy/Georgia).

Mareček, Martin and Skalsky, Martin (2005) *Zdroj* (The Source; Bionaut Films, Czech Republic).

Rivkin, Amanda (2011) *Baku-Tbilisi-Ceyhan Pipeline*, www.amandarivkin.com/#/baku-tbilisi-ceyhan-oil-pipeline/btc02 (accessed April 2013).

Documents, Reports and Publicity Material

Note: the vast majority of these documents can be downloaded from the websites of the organisations concerned. The archive of BTC project documents is accessible on the BP Caspian website, http://www.bp.com/subsection.do?categoryId=9006630&contentId=7013422 (accessed April 2013).

Amnesty International (2003) 'Human Rights on the Line: the Baku-Tbilisi-Ceyhan Pipeline Project, http://www.amnesty.org.uk/uploads/documents/doc_14538.pdf (accessed May 2013).

Baku-Ceyhan Campaign (2003a) 'Review of Project Environmental Impact Assessment – Executive Summary (Turkey Section)' http://www.bakuceyhan.org.uk/eia_review.htm (accessed April 2013)

Baku-Ceyhan Campaign (2003b) 'Review of Project Environmental Impact Assessment – 3. Consultation on the Project (Turkey Section)', http://www.thecornerhouse.org.uk/sites/thecornerhouse.org.uk/files/ (accessed May 2013).

Baku-Ceyhan Campaign (2003c) 'Evaluation of Compliance of the Baku-Tbilisi-Ceyhan (BTC) Pipeline with the Equator Principles, Supplementary Appendix to the EIA Review – BTC Pipeline (Turkey section)', http://www.bakuceyhan. org.uk/publications/Equator_Principles.pdf (accessed May 2013).

Baku-Ceyhan Campaign (2003d) 'Baku-Tbilisi-Ceyhan Project: IFC Response to Submissions Received During 120-day Comment Period for BTC Pipeline', Letter to International Finance Corporation from The Corner House, 3 November 2003, http://www.thecornerhouse.org.uk/sites/thecornerhouse.org.uk/files/ (accessed May 2013).

Baku-Ceyhan Campaign (2004a) 'Statement of Ferhat Kaya Regarding Arrest and Ill-treatment Following Work on BTC Pipeline', http://www.bakuceyhan.org.uk/ ferhat_statement.htm (accessed April 2013).

Baku-Ceyhan Campaign (2004b) 'Baku-Tbilisi-Ceyhan Oil Pipeline: Human Rights, Social and Environmental Impacts Turkey Section, Final Report of the Fourth Fact Finding Mission Ardahan and İmranlı Regions', 19–27 September.

Baku-Ceyhan Campaign (2005) 'Baku-Tbilisi-Ceyhan Oil Pipeline: Human Rights, Social and Environmental Impacts, Georgia and Turkey Sections, Report of Fact Finding Mission', 16–21 September, www.baku.org.uk (accessed May 2013).

Bank Information Center, CEE Bankwatch, Friends of the Earth US, Green Alternative, National Ecological Center of Ukraine, Platform (2003) 'Second International Fact-Finding Mission to Baku-Tbilisi-Ceyhan Pipeline, Georgia Section, Initial Summary Report', 4 June.

BBC (2005) 'Storyville – The Curse of Oil: Producers' Diary', previously available at http://www.bbc.co.uk/bbcfour/documentaries/storyville/oil-diary5.shtml

Borjomi-Kharagauli National Park (2002) 'Borjomi-Kharagauli National Park', Borjomi.

BP (2004) 'BP Azerbaijan, Sustainability Report', www.bp.com (accessed May 2013).

BP (2005) 'BP in Georgia, Sustainability Report', www.bp.com (accessed May 2013).

BP (2011) 'If Lightning Strikes' http://www.bp.com/ (accessed April 2013).

Bretton Woods Project (2006) 'Highlights of CAO Meeting with UK NGOs', Tuesday 28 November, http://brettonwoodsproject.org/art-547195 (accessed May 2013).

BTC (2003a) 'Breaking New Ground: Working with the Community to Enhance the Social Benefits of Oil Development and Export', March.

BTC (2003b) 'Safe, Silent and Unseen: Baku-Tbilsi-Ceyhan Pipeline Project'.

BTC (2003c) 'Between Two Seas: Baku-Tbilisi-Ceyhan Pipeline Project', February.

BTC (2004) 'Advertisement on Georgian Radio by Ed Johnson, BTC Manager, Georgia'.

BTC (2006) 'BTC Quick Facts'.

BTC (2009) 'Timeline of Notable BTC Milestones and Achievements'.

BTC/CDAP (2003) 'Interim Report on Azerbaijan and Georgia', August.

BTC/CDAP (2004) 'Letter to Sir John Browne', 30 December.

BTC/CDAP (2007) 'Final Report and Conclusions'.

BTC/EIA (2003) 'BTC Project EIA Turkey, Update to Public Consultation and Disclosure Plan – Final EIA'.

BTC/ESAP (2002a) 'Environment and Social Action Plan'

BTC/ESAP (2002b) 'Environment and Social Action Plan, Environment and Social Commitments Register – Georgia'.

BTC/ESAP (2002c) 'Environment and Social Action Plan, Environment and Social Commitments Register – Azerbaijan'.

BTC/ESAP (2003) 'Contractor Control Plan, Transport Management, Georgia'.

BTC/ESAR (2008) 'Environmental and Social Annual Report (Operations Phase)'.

BTC/ESIA (2002a) 'BTC Project ESIA, Executive Summary'

BTC/ESIA (2002b) 'BTC Project ESIA Azerbaijan, Final ESIA'.

BTC/ESIA (2002c) 'BTC Project ESIA Georgia, Final ESIA'.

BTC/ESIA (2002d) 'BTC Pipeline ESIA Georgia, Final ESIA, Social Impacts and Mitigation'.

BTC/ESIA (2002e) 'BTC Project ESIA Georgia – Appendix D Social Baseline – Annex 1 Community Survey Summary'.

BTC/ESIA (2002f) 'BTC/SCP Project ESIA Georgia, Appendix E Environmental Impacts and Mitigation Annex III Cultural Heritage Management Plan'.

BTC/ESIA (2002g) 'BTC Pipeline ESIA Azerbaijan, Final ESIA, Geohazards'.

BTC/ESIA (2002h) 'BTC Pipeline ESIA Azerbaijan, Regulatory Review of Environmental and Social Issues'.

BTC/ESIA (2002i) 'BTC Project ESIA Georgia Response to Comments (Public Information Disclosure)'.

BTC/ESIA (2002j) 'BTC Project ESIA Georgia, Final ESIA, Project Alternatives'.

BTC/ESIA (2002k) 'BTC Project ESIA Azerbaijan, Environmental and Social Impact Assessment Methodology'.

BTC/ESIA (2003) ESIA – 'Supplementary Lenders Information Part C, Appendix II: Public Consultation and Disclosure Plan – BTC and SCP Projects Georgia'.

BTC/ESIA (2004) 'Appendix 1: Georgia ESIA Continuing Activities'.

BTC/ESM (2007) 'BTC Environment and Social Management Plan – Employment and Training, Azerbaijan and Georgia Operations'.

BTC/ESR (2004a) 'Fourth Quarter Environmental and Social Report to Lenders'.

BTC/ESR (2004b) 'BTC Project Environmental and Social Annual Report'.

BTC/Georgia (2004) 'Agreement Between BTC and the Government of Georgia on the Establishment of a Grant Program for Georgia', 19 October.

BTC/IEC (2004) 'Report of the Post-Financial Close Independent Environmental Consultant, First Site Visit', February–March.

BTC/IEC (2006) 'Report of the Post-Financial Close Independent Environmental Consultant, Eighth Site Visit', October.

BTC/IEC (2010) 'Report of the Post-Financial Close Independent Environmental Consultant, Twelfth Site Visit', July.

BTC/OSR (2005) 'Oil Spill Response Plan, BTC Operations in Georgia, Appendix 12'.

BTC/PCIP (2003) 'BTC Project Community Investment Plan'.

BTC/PMDI (2006) 'BTC Oil Pipeline Oil Spill Response Team Audit Report, Pipeline Monitoring and Dialogue Initiative (PMDI) Program II Phase'.

BTC/RAP (2002a) 'Resettlement Action Plan Georgia – Annex 4 List of Consulted Villages'.

BTC/RAP (2002b) 'Overview of Project Affected Populations, Resettlement Action Plan – Georgia'.

BTC/RAP (2002c) 'Land Acquisitions and Procedures and Implementing Responsibilities – Georgia'.

BTC/RAP (2002d) 'BTC Project Resettlement Action Plan, Azerbaijan'.

BTC/RAP (2002e) 'Introduction, BTC Project Resettlement Action Plan, Georgia'.

BTC/RAP (2002f) 'Georgia Costs and Budgets'.

BTC/RAP (2002g) 'Legislative, Regulatory and Legal Framework, Resettlement Action Plan, Georgia'.

BTC/RAP (2003) 'RAP – Resettlement Action Plan Final, Part A Overview'.

BTC/RR (2003) 'Regional Review: Economic, Social and Environmental Overview of the Southern Caspian Oil and Gas Projects'.

BTC/SCP/CDI (2011) 'Community Development Initiative Report for April–June 2011'.

BTC/SR (2005) 'BTC in Georgia, Sustainability Report'.

BTC/SRAP (2003a) 'Overview of SRAP Panel Review Findings and Monitoring Terms of Reference'.

BTC/SRAP (2003b) 'Georgia SRAP Review Findings', August.

BTC/SRAP (2003c) 'Azerbaijan SRAP Review Findings', August.

BTC/SRAP (2004a) 'Georgia SRAP Panel Review', February.

BTC/SRAP (2004b) 'Georgia SRAP Panel Review', July.

BTC/SRAP (2004c) 'Azerbaijan SRAP Panel Review', February.

BTC/SRAP (2005a) 'Georgia SRAP Panel Review', March.

BTC/SRAP (2005b) 'Georgia SRAP Review Findings', September.

BTC/SRAP (2006a) 'Georgia SRAP Review Findings', March.

BTC/SRAP (2006b) 'Turkey SRAP Review Findings', March.

BTC/SRAP (2007) 'Georgia SRAP Review Findings', June.

BTC/SRAP (2008) 'Georgia SRAP Panel Review', April.

Campagna per la Riforma della Banca Mondiale, Kurdish Human Rights Project, Platform, The Corner House (2003) 'International Fact-Finding Mission: Baku-Tbilisi-Ceyhan Pipeline – Turkey section, June'.

Caspian Revenue Watch (2003) *Caspian Oil Windfalls: Who Will Benefit?*, New York: Open Society Institute.

CEE Bankwatch (2003) 'Azerbaijan-Georgia-Turkey Pipeline Systems: An Evaluation of the Public Disclosure and Consultation Process in Azerbaijan', Prague, http://bankwatch.org/documents/agt_pipeline_systems_10_03.pdf (accessed May 2013).

CEE Bankwatch (2004) 'Public Participation, Access to Information and Development Banks in the UN ECE region', 1 June, www.unece.org/env/pp/ppif/ppif-ceebankwatch.06-01-04.pdf (accessed April 2013).

CEE Bankwatch (2006) 'Beyond the Pale: Myths and Realities about the BTC Development Model', Prague, http://bankwatch.org/documents/BTC_development_model.pdf (accessed May 2013).

CEE Bankwatch (2008) 'The EBRD Independent Recourse Mechanism Review: CEE Bankwatch's Comments on the Existing IRM, http://bankwatch.org/documents/bwn_comments_IRM_06_08_FINAL.pdf (accessed May 2013).

CEE Bankwatch, Green Alternative, Georgian Young Lawyers Association (2003) 'Baku-Tbilisi-Ceyhan Pipeline: Review of Land Acquisition and Compensation Process', Tbilisi.

CEE Bankwatch, Friends of the Earth, Green Alternative, Les Amis de la Terre, National Ecological Centre of Ukraine (2004) 'Third Fact Finding Mission: Azerbaijan, Turkey and Georgia Pipeline Project: Georgia Section', October, http://www.bakuceyhan.org.uk/publications/ffm_georg_report_04.pdf (accessed May 2013).

Centre for Civic Initiatives, CEE Bankwatch, Green Alternative, Friends of the Earth, Les Amis de la Terre (2004) 'Third Fact Finding Mission Report Azerbaijan-Georgia-Turkey, Azerbaijan Section', October, http://www.bakuceyhan.org.uk/publications/ffm_azeri_report_04.pdf (accessed May 2013).

Centre for Civic Initiatives, CEE Bankwatch, Committee for the Protection of Oil Workers' Rights, Green Alternative, Kurdish Human Rights Project, Platform, Urgewald (2005a) 'Baku-Tbilisi-Ceyhan Oil Pipeline: Human Rights, Social and Environmental Impacts, Georgia Section, Final Report of Fourth Fact Finding Mission, Tetriskaro, Borjomi and Akhaltsikhe Regions', 16–18 September, http://www.bakuceyhan.org.uk/publications/FFM05Georgia.pdf (accessed May 2013).

Centre for Civic Initiatives, CEE Bankwatch, Committee for the Protection of Oil Workers' Rights, Green Alternative, Kurdish Human Rights Project, Platform, Urgewald (2005b) 'Baku-Tbilisi-Ceyhan Oil Pipeline: Human Rights, Social and Environmental Impacts, Turkey Section, Final Report of the Fact Finding Mission', 18–21 September, http://www.bakuceyhan.org.uk/publications/FFM05turkey.pdf (accessed May 2013).

Coffey International Development (2009) 'Validation of the Extractive Industries Transparency Initiative in the Republic of Azererbaijan', http://eiti.org/files/Azerbaijan%20EITI%20Validation%20Report.pdf (accessed April 2013).

Cooke, K. (2006) 'Power Games in the Caucasus', http://news.bbc.co.uk/1/hi/business/4964316.stm (accessed April 2013).

The Corner House (2011) 'The Baku-Tbilisi-Ceyhan (BTC) Oil Pipeline', http://www.thecornerhouse.org.uk/background/baku-tbilisi-ceyhan-btc-oil-pipeline (accessed April 2013).

The Corner House, Kurdish Human Rights Project, Platform (2011) 'Company Undertakings on the OECD Guidelines and the Implications of the UK National Contact Point's March 2011 Final Statement on the BTC Specific Instance', http://www.thecornerhouse.org.uk/sites/thecornerhouse.org.uk/files/Implications%20of%20NCP%20Final%20Statement%20BP_1.pdf (accessed April 2013).

The Corner House, Kurdish Human Rights Project, Platform, CEE Bankwatch, Friends of the Earth (2008) 'Export Credits Guarantee Department and Sustainable Development', Memorandum to the House of Commons Environmental Audit Committee, http://bankwatch.org/documents/EAC_BTC_Conflict_26.8.08.pdf (accessed April 2013).

Crude Accountability (2012) 'After the BTC Pipeline and EITI Validation: Where are Prosperity and Transparency in Azerbaijan?', http://www.osce.org/odihr/94979 (accessed April 2013).

Decker, K., Hlobil, P., Kochkladze, M., Marriot, J., Nagel, W., Urbansky, Y. and Welch, C. (2003) 'Letter to the Executive Directors of the IFC', 14 August, http://www.baku.org.uk/correspondence.htm (accessed April 2013).

EBRD (2004) 'Address by Noreen Doyle at the Ceremony to Mark the Financing of the BTC Pipeline', Baku, 3 February.

EBRD/IRM (2009) 'Problem-solving Completion Report (PsCR) Complaint: BTC Georgia/Atskuri Village, Georgia', http://www.ebrd.com/downloads/integrity/0809pscr.pdf (accessed April 2013).

EEC (1990) 'Council Directive 90/313/EEC of 7 June 1990 on the freedom of access to information on the environment', *Official Journal L* 158, 56–58.

EITI (2003) 'Principles', http://eiti.org/eiti/principles (accessed April 2013).

EITI (2006) 'Validation Guide', London: DFID.

EITI (2012a) 'Azerbaijan – In-kind Payments Larger Than Cash', http://www.eiti.org, 20 June (accessed April 2013).

EITI (2012b) 'Nigeria EITI: Making Transparency Count, Uncovering Billions', EITI case study 20 January.

Equator Principles (2003) 'Equator Principles: Environmental and Social Risk Management for Project Finance', http://www.equator-principles.com/ (accessed April 2013).

Friends of the Earth (2003) 'Conflict, Corruption, and Climate Change', http://www.foe.co.uk/resource/briefings/conflict_climate_change.pdf (accessed April 2013).

Georgia (2002) 'Continuing Activities Under the Environmental Permit for the BTC ESIA', Ministry of Environment, 30 November (reproduced in BTC/ESIA (2004) 'Appendix 1: Georgia ESIA Continuing Activities')

Georgia (2004) *'Law of Georgia: On Environmental Protection Permit', in Environmental Legislation of the South Caucasus Countries volume 1,* Tbilisi: Caucasus Environmental Network.

Gillard, M. (2004) 'Memorandum', Written Evidence to the House of Commons Trade and Industry Select Committee, http://www.publications.parliament.uk/pa/cm200405/cmselect/cmtrdind/374/374we07.htm (accessed April 2013).

Global Security (2003) 'GTEP – A Unique Mission for the Corps', http://www.globalsecurity.org/military/library/news/2003/01/mil-030122-usmc01.htm (accessed April 2013).

Green Alternative (2002) 'Analysis of Agreement among the Azerbaijan Republic, Georgia and the Republic of Turkey Relating to the Transportation of Petroleum via the Territories of The Azerbaijan Republic, Georgia and the Republic of Turkey through the Baku-Tbilisi-Ceyhan Main Export Pipeline', Tbilisi, http://bankwatch.org/documents/analysis_btc_gaoxfam_02.pdf (accessed May 2013).

Green Alternative (2003) 'Report on BTC Construction Compliance', Tbilisi: Green Alternative.

Green Alternative (2004a) 'Letter of Complaint to Compliance Advisor/ Ombudsman', 21 May, http://bankwatch.org/documents/complaint_ga_ifc_05_04.pdf (accessed April 2013).

Green Alternative (2004b) 'Complaint to IFC Ombudsman', http://www.cao-ombudsman.org/html-english/documents/Tsemi2AssessmentReport_FINAL.pdf.

Green Alternative (2005) 'BTC Pipeline: An IFI Recipe for Increasing Poverty', http://bankwatch.org/sites/default/files/report_btc_poverty_10_05.pdf (accessed May 2013).

Green Alternative/CEE Bankwatch (2005) 'BTC Pipeline – an IFI Recipe for Increasing Poverty', http://bankwatch.org/documents/report_btc_poverty_10_05.pdf (accessed May 2013).

Green Alternative, Georgian Young Lawyers Association, CEE Bankwatch Network (2004) 'Baku-Tbilisi-Ceyhan Pipeline, the BTC Company and Social and Environmental Protection Obligations', Tbilisi, May.

Green Alternative, CEE Bankwatch Network, Campagna per la Riforma della Banca Mondiale, Platform, Friends of the Earth US, Bank Information Center (2002) 'International Fact Finding Mission, Preliminary Report, Azerbaijan, Georgia, Turkey Pipelines Project: Georgia Section'.

Greenpeace (2012) 'Stop Esso', http://www.greenpeace.org.uk/climate/stop-esso (accessed April 2013).

GYLA (2003) 'Baku-Tbilisi-Ceyhan: The Legal Analysis of the Host Government Agreement', Tbilisi: Georgian Young Lawyers Association.

Hildyard, N. and Muttit, G. (2006) 'Turbo-charging Investor Sovereignty: Investor Agreements and Corporate Colonialism', http://www.thecornerhouse.org.uk/sites/thecornerhouse.org.uk/files/HGAPSA.pdf (accessed April 2013).

Host Government Agreement (2000a) 'Host Government Agreement Between and Among the Government of the Azerbaijan Republic and the State Oil Company of the Azerbaijan Republic, BP Exploration (Caspian Sea) Ltd, Statoil BTC Caspian AS, Ramco Hazar Energy Limited, Turkiye Petrolleri A.O., Unocal BTC Pipeline Ltd, Itochu Oil Exploration (Azerbaijan) Inc., Delta Hess (BTC) Limited', http://subsites.bp.com/caspian/BTC/Eng/agmt1/agmt1.PDF (accessed May 2013).

Host Government Agreement (2000b) 'Host Government Agreement Between and Among the Government of Georgia and the MEP Participants', http://subsites.bp.com/caspian/BTC/Eng/agmt2/agmt2.PDF (accessed May 2013).

Host Government Agreement (2000c) 'Host Government Agreement Between and Among the Government of Turkey and the MEP Participants', http://subsites.bp.com/caspian/BTC/Eng/agmt3/agmt3.PDF (accessed May 2013).

House of Commons (2004) Memoranda Regarding ECGD Support for the Baku-Tbilisi-Ceyhan Pipeline Project, Trade and Industry Select Committee, London: HMSO.

House of Commons (2005a) Implementation of ECGD's Business Principles: Government Response to the Committee's Ninth Report of Session 2004–5, HC-374-I, London: HMSO.

House of Commons (2005b) Implementation of ECGD's Business Principles: Ninth Report of Session 2004–5, vol. 2, Oral and Written Evidence, HC-274-II, London: HMSO.

House of Commons (2009) Russia: A New Confrontation?: Tenth Report of Session 2008–9, Defence Select Committee Report together with formal minutes, oral and written evidence, HC 987, London: HMSO.

House of Lords (1995) Lords Hansard, 12 June, London: HMSO.

Hume, I. (1995) 'Effects of Road Traffic Vibration on Historic Buildings', Context, 47, www.ihbc.org.uk (accessed April 2013).

IFC (nd) 'Fact Sheet on Chad-Cameroon Pipeline', www.ifc.org, (accessed April 2013).

IFC (2003a) 'BTC Pipeline and ATC Phase 1 Projects, Environmental and Social Documentation, IFC Responses to Submissions Received During the 120 Day Comment Period', www.ifc.org (accessed April 2013).

IFC (2003b) 'IFC Invests in Caspian Oil and Pipeline Projects', 5 November, www.worldbank.org.tr (accessed April 2013).

IFC (2003c) 'BTC Pipeline Project: Review of Issues Concerning Horizontal Directional Drilling in Azerbaijan', Letter from Yasmin Tayab to Carol Welch (Friends of the Earth, USA) and Karen Decker (Bank Information Center), 16 September, http://www.bakuceyhan.org.uk/correspondence.htm (accessed May 2013).

IFC (2006) 'External Monitoring of the Chad-Cameroon Pipeline Project', *Lessons of Experience*, 1, September.

IFC/CAO (2004a) 'Seven Complaints Regarding the Baku-Tbilisi-Ceyhan (BTC) Pipeline Project Bashkovi, Dgvari, Rustavi, Sagrasheni, Tetriskaro and Tsikhisjvari, Georgia', http://www.cao-ombudsman.org/cases/document-links/documents/BTC (accessed May 2013).

IFC/CAO (2004b) 'Assessment Report: Complaint Regarding the Baku-Tbilisi-Ceyhan (BTC) Pipeline Project, Rustavi, Georgia', Washington, DC: International Finance Corporation.

IFC/CAO (2008) 'BTC Pipeline: Summary of Complaints 2003 to March 2008', Washington, DC: International Finance Corporation.

IFC/EBRD (2003) 'Report of the IFC and EBRD Multi-Stakeholder Forum (MSF) Meetings on the Baku-Tbilisi-Ceyhan Oil Pipeline, ACG Phase 1, Shah Deniz and South Caucasus Pipeline Projects', http://www.ebrd.com/pages/project/eia/18806msf.pdf (accessed May 2013).

Intergovernmental Agreement (1999) 'Agreement Among the Azerbaijan Republic, Georgia and the Republic of Turkey relating to the Transportation of Petroleum via the Territories of the Azerbaijan Republic, Georgia and the Republic of Turkey through the Baku-Tbilisi-Ceyhan Main Export Pipeline', 18 November, http://subsites.bp.com/caspian/BTC/Eng/agmt4/agmt4.PDF (accessed May 2013).

International Alert (2004) 'Oil and the Search for Peace in the South Caucasus: The Baku-Tbilisi-Ceyhan (BTC) Oil Pipeline', September, London: International Alert.

Jeter, J. (2005) 'Technical and Geo-political Challenges Facing Onshore Gas and Oil Pipelines and their Influence on Routing', in M. Sweeney (ed.) *Terrain and Geohazard Challenges Facing Onshore Oil and Gas Pipelines*, London: Thomas Telford, 1–7.

Jones, E. (2001) 'US Caspian Energy Diplomacy: What has Changed?', Caspian Studies Program, Harvard University, 11 April.

KHRP (2003) 'Campaigners Urge Moratorium on Controversial BP Pipeline', http://www.khrp.org/khrp-news/news-archive/2003-news/108-campaigners-urge-moratorium-on-controversial-bp-pipeline.html (accessed April 2013).

Mansley, M. (2003) *'Building Tomorrow's Crisis – The Baku-Tbilisi-Ceyhan and BP: A Financial Analysis'*, London: Platform.

Marriott, J. (2003) 'The Hive and the Hard Drive', *New Internationalist*, http://newint.org/features/2003/10/01/economics/ (accessed May 2013).

Marriott, J. (2005) 'The Mapping of the Pipeline in the Metropolis', in A. Franke (ed.) *B-Zone: Becoming Europe and Beyond*, Berlin: Kunst-Werke, 96–99.

Marriott, J. (2007) 'The End of the Empire of Gog and Magog', *Soundings*, 35, 76–91.

Marriott, J. and Minio-Paluello, M. (2012) *The Oil Road: A Journey to the Heart of the Energy Economy*, London: Verso.

Marriott, J. and Muttit, G. (2006) 'BP's Baku-Tbilisi-Ceyhan Pipeline: The New Corporate Colonialism', *New Solutions: A Journal of Environmental and Occupational Health Policy*, 16, 1, 21–63.

Meacher, M. (2005) 'Casualities of the Oil Stampede', *Guardian*, 15 June.

Moore Stephens (2007) 'The Committee on the Extractive Industries Transparency Initiative of the Republic of Azerbaijan, Independent Accountants Report for the Year Ended 31 December 2006', www.eiti.org (accessed April 2013).

Moser, P (2003) 'In the Matter of the Baku-Tbilisi-Ceyhan Pipeline: Counsel's Opinion', http://www.bakuceyhan.org.uk/publications/opinion_moser.pdf (accessed May 2013).

NCEIA (2002) 'Advisory Review of the Environmental and Social Impact Assessment Reports of the Baku-Tbilisi-Ceyhan Oil Pipeline and the South Caucasus Gas Pipeline in Georgia', 2nd edition, 22 November, Utrecht: NCEIA.

NCEIA (2003) 'Advisory Review of the Environmental and Social Impact Assessment Reports and Supplementary Information of the Baku-Tbilisi-Ceyhan Oil Pipeline and South Caucasus Gas Pipeline in Georgia', 15 October, Utrecht: NCEIA.

NCEIA (2004) 'Advisory Review of the Compliance of Project Implementation with the Environment for the Baku-Tbilisi-Ceyhan Oil Pipeline and the South Caucasus Gas Pipeline in Georgia', https://zoek.officielebekendmakingen.nl/kst-26234-41-b2.pdf (accessed May 2013).

OECD (2000) 'The OECD Guidelines for Multinational Enterprises', Paris: OECD http://www.oecd.org/dataoecd/56/36/1922428.pdf (accessed April 2013).

OSCE (2003) 'Election Observation Mission Report', http://www.osce.org/odihr/elections/azerbaijan/13467 (acccesed May 2013).

OWRP (2004) 'Oil Workers' Rights Protection Organisation Results of Monitoring on Impact of Baku-Tbilisi-Ceyhan Oil Pipelines to Solution of Social-economic Problems in Azerbaijan Districts', http://www.nhmt-az.org/ts_general/eng/prj/prj5-7.htm (accessed April 2013).

Pagnamenta, R. (2005) 'A Lifeline, But Not for Them', New Statesman, 21 November.

Platform (2006) 'Platform', http://old.platformlondon.org/bodypolitic.asp (accessed May 2013).

Platform, The Corner House, Friends of the Earth, Compana per la Riforma della Banca Mondiale, CEE Bankwatch, KHRP (2003) Some Common Concerns: Imagining BP's Azerbaijan-Georgia-Turkey Pipeline System, London.

Rising Tide (2002) 'London Rising Tide Invades ERM over BP Baku Pipeline: A Report From this Monday's Occupation of ERM', http://risingtide.org.uk/content/london-rising-tide-invades-erm-offices-over-bp-baku-pipeline (accessed April 2013).

Rivkin, A. (2011) 'Shaken by Change', National Geographic, 8 June.

Rondeaux, C. (2004) 'A Pipeline to Promise, or a Pipeline to Peril', International Reporting Project, 1 December, http://www.internationalreportingproject.org/stories/view/a-pipeline-to-promise-or-a-pipeline-to-peril (accessed April 2013).

Shell (nd) 'Brent Spar – The Dossier', http://www.static.shell.com/static/gbr/downloads/e_and_p/brent_spar_dossier.pdf (accessed April 2013).

Shiston, D., Lee, E., Pollos-Pirallo, S., Morgan, D., Clarke, J., Fookes, P. and Brunsden, D. (2005) 'Terrain Evaluation and Site Investigations for Design of the Trans-Caucasus Oil and Gas Pipelines in Georgia', in M. Sweeney (ed.) Terrain and Geohazard Challenges Facing Onshore Oil and Gas Pipelines, London: Telford, 289–300.

Spence, S. (2004) 'Keeping the Promises – A Management Framework for Delivering ESIA Commitments During Pipeline Construction', SPE International Conference on Health, Safety, and Environment in Oil and Gas Exploration and Production, 29–31 March 2004, Calgary, Alberta, Canada.

Tatashidze, Z., Tsereteli, E., Gaprindashvili, M., Tsereteli, N., and Taliashvili, D. (2006), 'Dgvari Land-Slide (Borjomi Region) and the Risk of its Implication on Baku-Tbilisi-Ceyhan Pipeline Functioning', *Bulletin of the Georgian National Academy of Sciences*, 304–306.

Tate (2012) 'APG – Artists Placement Group', http://www2.tate.org.uk/ artistplacementgroup/ (accessed April 2013).

Transparency International (2009) 'Corruption Perception Index', http://archive. transparency.org/policy_research/surveys_indices/cpi/2009/cpi_2009_table (accessed April 2013).

Transport and Road Research Laboratory (TRRL) (1990) 'Traffic Induced Vibrations in Buildings', Report 246, Wokingham: TRRL.

UKNCP (2011) 'OECD Guidelines for Multinational Enterprises – Specific Instance: The BTC Pipeline', revised final statement, 21 February, http://www.bis.gov.uk/ assets/biscore/business-sectors/docs/r/11-766-revised-final-statement-ncp-btc (accessed April 2013).

UN (2007) 'Declaration on the Rights of Indigenous Peoples', http://social.un.org/ index/IndigenousPeoples/DeclarationontheRightsofIndigenousPeoples.aspx (accessed April 2013).

UNECE (1998) 'Convention on Access to Information, Public Participation in Decision and Access to Justice in Environmental Matters', Åarhus, Denmark, 25 June.

World Bank (1999) 'World Bank Operational Manual: OP 4.01 Environmental Assessment and Successor Documents', www.worldbank.org (accessed May 2013).

World Bank (2000) 'GP 14.70 Involving Non-Governmental Organisations in Bank Supported Activities', www.worldbank.org (accessed May 2013).

World Bank (2001) *Georgia – Energy Transit Institution Building Project*. Washington, DC: World Bank. http://documents.worldbank.org/curated/en/2001/02/1000517/ georgia-energy-transit-institution-building-project (accessed May 2013).

World Bank (2008) *Implementing the Extractive Industries Transparency Initiative*, Washington, DC: The World Bank.

WWF (2003) 'Disclosure Period on ESIA Documentation for Proposed BTC Pipeline: WWF Comment on BTC Application to EBRD for Finance', 13 October.

Index

Note: page numbers in italics refer to figures; those in bold to tables. 'n' denotes notes with the note number indicated by the number following the 'n'.

Material Politics: Disputes Along the Pipeline, First Edition. Andrew Barry.
© 2013 John Wiley & Sons, Ltd. Published 2013 by John Wiley & Sons, Ltd.

Printed in the United States
by Bookmasters

Printed in the United States
By Bookmasters